KEEPERS OF THE FLAME

Keepers of the Flame

THE ROLE OF FIRE IN
AMERICAN CULTURE
1775–1925

MARGARET HINDLE HAZEN
AND
ROBERT M. HAZEN

PRINCETON UNIVERSITY PRESS, PRINCETON, N.J.

Copyright © 1992 by Princeton University Press
Published by Princeton University Press, 41 William Street,
Princeton, New Jersey 08540

All Rights Reserved

Library of Congress Cataloging-in-Publication Data

Hazen, Margaret Hindle.
Keepers of the flame : the role of fire in American
culture, 1775–1925 / Margaret Hindle Hazen
and Robert M. Hazen.
p.   cm.
Includes bibliographical references and index.
ISBN 0-691-04809-6
1. United States—Social life and customs.
2. Fire—Social aspects—United States.
I. Hazen, Robert M., 1948– .   II. Title.
E161.H39 1992
973.5—dc20   92-12469

This book has been composed in Adobe Goudy
Designed by Jan Lilly

Princeton University Press books are printed on
acid-free paper, and meet the guidelines for permanence
and durability of the Committee on Production Guidelines
for Book Longevity of the Council
on Library Resources

Printed in the United States of America

2   4   6   8   10   9   7   5   3   1

For Ben and Elizabeth

Contents

Preface

Energy is essential to life and all its pursuits. Its acquisition and use are inevitably linked to a society's structure, values, and dreams. For Americans before World War I, fire provided the energy for a broad range of endeavors, filling homes with light, warmth, and a means to prepare food, and driving American mines, forges, factories, and transportation. Fire was everywhere, and almost everyone faced the open flame every day.

No single volume could review all aspects of fire in America. We have attempted instead a broad approach to the subject, fully realizing that almost any paragraph could be expanded into a dissertation. Indeed, some fire topics have been analyzed in depth already. The more spectacular conflagrations of American history, as well as the firefighting fraternity and its distinctive accoutrements, are featured in numerous volumes and museum exhibits. Distinctive fire-related objects—particularly collectible cooking and lighting devices—have also received considerable attention. Yet, the larger significance of fire in American culture and daily life has attracted less study.

Our chapter organization reflects our aim to provide this overview by identifying themes that focus on a specific view of fire. Following a brief introductory chapter that places fire in its American context, we consider the paradoxical nature of fire: the valued servant of many uses (chapter 2) versus the dreaded master of destruction (chapter 3). Chapter 4 details some of the efforts devised over 150 years to extinguish unwanted fire and counteract its destructiveness. The focus narrows in the next two chapters. Chapter 5 considers the everyday problems of fire use and explores the theme of the gradual insulation of most users from the open flame. Chapter 6 looks at the quest for a scientific understanding of the principles of combustion, while chapter 7 highlights the simultaneous drive to preserve the nostalgic interpretation of the flame. Finally, in chapter 8, we examine fire from our twentieth-century perspective.

We have consciously omitted three aspects of fire. Explosives, weapons, and the use of fire in warfare are not considered here. Likewise, we refer only in passing to fire symbolism in organized religion and to the fire practices of Native Americans. We confine ourselves, therefore, to the peacetime—though not always peaceful—pursuits of secular America from the time of the Revolution to the third decade of the twentieth century, the point at which electricity had finally conquered urban America and had just begun to spread to rural areas.

We have benefited from the help and advice of many individuals. Special thanks are due to our friends and colleagues at the National Museum of American History, Smithsonian Institution. In particular, Ann Serio of the Division of Domestic Life and John Fleckner and David Haberstich of the Archives Center provided valuable assistance.

We thank Charles T. Prewitt, Director of the Geophysical Laboratory of the Carnegie Institution of Washington, for his continued support and encouragement. Partial financial support for obtaining and reproducing illustrations was provided by the Carnegie Institution of Washington and by George Mason University. The book would not have been possible without the support and advice of our editor, Edward Tenner. His perceptive reading of the early drafts and his thorough understanding of the subject matter helped enormously as we prepared the final manuscript. Inevitably, his viewpoint helped us to sharpen our own. The book also benefited from the careful copyediting of Gretchen Oberfranc.

In addition, we thank our families and friends, whose experiences and anecdotes related to fire in America provided a continuing source of inspiration and encouragement.

KEEPERS OF THE FLAME

Heritage of Flame

The Europeans who colonized America during the seventeenth century arrived with optimism, determination, and fire. Unlike the fire-worshiping empire builders of antiquity, however, these no-nonsense American settlers mostly ignored ceremony as they passed the torch to the New World. There was work to be done, and fire provided the most potent and reliable energy source to accomplish that work.

These migrating fire builders could hardly have picked a better starting point for their endeavors than the east coast of North America. "Here is good living for those that love good Fires," exulted Francis Higginson in 1630. An Englishman accustomed to fuel shortages, Higginson could barely contain his enthusiasm as he eyed the seemingly endless forests of New England: "Though it bee here somewhat cold in the winter, yet here we have plenty of Fire to warme us, and that a great deale cheaper then they sel Billets and Faggots in *London*." Even these claims apparently fell short of the glorious truth, for the writer quickly corrected himself. "Nay, all *Europe*," he stated categorically, "is not able to afford so great Fires as *New-England*."[1]

Despite increases in population and a corresponding decrease in fuel supplies over the years, this early fascination with "great Fires" endured. Indeed, fire power, intensifying with each new wave of immigration, ultimately transformed the continent. Colonists harnessed the energy of the flame to cook their food, heat their homes, and light their way at night. Following the example of the indigenous peoples, they also hunted game and cleared the land with fire. Early iron- and glassworkers relied on the energy of fire, as did coopers, tanners, tinsmiths, dyers, chemists, shipwrights, and most other craftsmen. Tobacco, the money crop that saved Virginia, depended on fire both for curing the leaves and for releasing the benefits of the weed.

This is not to say that early Americans rejected other energy sources as they scrambled to dominate the New World. In addition to chemical energy, in the form of animal and human muscle power, colonists harnessed waterpower to operate mills and to work the bellows at ironworks. Wind power, which had transported the newcomers to America in the first place,

Fire is a central element in "Among the Pines—A First Settlement" by Currier and Ives. American settlers depended on fire as their principal energy source.

continued to propel boats and, more rarely, activated pumps, mills, and the occasional railroad car.

Nevertheless, it was fire—that mystical, magical tool of antiquity—that energized the majority of domestic and commercial activities during the colonial period and that by the nineteenth century had become a leitmotif of the country's life. "For, Lo! We live in an Iron Age— / In the age of Steam and Fire!" wrote a poet who was mesmerized by the fire-powered steam engines that by the 1850s had begun to revolutionize American transportation, agriculture, and industry.[2] Other types of fires animated diverse sectors of society simultaneously. Gas fires glowed in the cities, campfires dotted the prairie, and blacksmiths' coals burned brightly in almost every settlement between. Counterbalancing these restless flames of progress, meanwhile, were legions of "home fires" that, via the fireplace, stove, candle, and lamp, provided the necessary energy to run individual households and nurture domestic life. Americans of the nineteenth century liked to measure their national progress according to a variety of standards—tons of iron and steel produced, miles of railroad tracks laid, and even the number of brass bands "discoursing sweet music." At the foundation of it all, however, lay the proven ability of the population at large to build and manipulate fires.

American factories and railroads were powered by fire. J. Frank Waldo's 1883 painting of the Clark Match Works in Oshkosh, Wisconsin, incorporates a variety of fire-related themes. Courtesy of the Oshkosh Public Museum.

The appeal of home fires provided powerful imagery for advertising. The Florence Machine Company incorporated an oil lamp and oil stove in this domestic scene from a trade card, ca. 1875.

The stove figures prominently in this view of an office, ca. 1895.

And the population at large is precisely who was involved. "We all keep Fires," noted Philip Vickers Fithian in a diary entry that referred specifically to the colonial Virginia family with whom he was living but that could have just as easily applied to Americans everywhere.[3] Only rarely does the world of the merchant resemble that of the mechanic, but in the realm of energy production in the pre-electric era, American society had a powerful leveler. Whatever a person's station or sphere—man or woman, adult or child, master or servant—fire was a perpetual fact of life.

The journalistic jottings of generations of Americans give some idea of the extraordinary dimensions of this reality:

Fire looks & feels most welcome; and I observe it makes our children remarkably garrulous & noisy. [1774]

This morning Michael Mick, Jr., put fire in the Furnace. [1810]

Medley and Ella froze themselves playing and singing in cold front room all the songs they knew. I enjoyed it all by the kitchen fire. [1867]

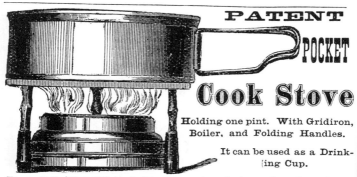

Everyone used fire. This portable "pocket cook stove" was offered for
"Travelers, Tourists, Camp Meetings, Family Use, and the Nursery; also
those who take their meals at their place of business." Advertisement from
Demorest's Illustrated Monthly Magazine, December 1875.

Went out foraging for firewood, & got enough dry sage brush to start
up the fire. [1902][4]

Fire was such a constant and central concern for most people that it functioned as the man-made counterpart of nature's sun. The day started and
ended at the hearth, and much of the work of the intervening hours re-
volved around the open flame.

On one level, fire maintenance constituted nothing more than a fairly
routine, if profoundly time-consuming, chore. Starting a fire could, under
ideal circumstances, be amazingly simple. Fires could be built almost anywhere with a bewildering variety of fuels. Both portable and—if properly
banked—preservable, fire permitted its users great freedom of mobility as
well as limited laxity in technique (one could always borrow a coal from a
neighbor). With the introduction of friction matches after 1827, it became
possible to start fires almost instantly whenever and wherever needed.

At the same time, however, fire maintenance was a deadly serious business. Even the slightest lapse in caution could precipitate the disastrous

transformation of fire from friendly ally to ravaging destroyer. The experience of Mrs. Orsemus Boyd, an army wife stationed in a remote area of Nevada during the late 1860s, illustrates the problem in stark detail. "Home" for the Boyds consisted of the basics: two tents, a bed, and, because winters were brutal, several heating devices. Specifically, "a large stove had been placed in the outer tent, and a huge fireplace built in the inner one. A large pine bunk, forming a double bed, occupied nearly all the spare space, and left only just room enough in front of the fire to seat one's self, and also to accommodate the tiniest shelf for toilet purposes." In such cramped quarters, as Mrs. Boyd explained, it was only a matter of time before fire the benefactor became fire the destroyer:

> In our inability to find suitable places for necessary articles, we were apt to use most inappropriate ones. On the occasion referred to, a lighted candle had been placed on the bed, where my husband seated himself without noticing the candle. Soon arose the accustomed smell of burning, and I executed my usual maneuver of turning about in front of the fire to see if my draperies had caught. The odor of burning continued to increase, yet I could find no occasion for it.
>
> The cause, however, was discovered when I leaned over the bed, and saw that a large hole had been burned in the center of Mr. Boyd's only uniform coat. He had been too intent on shielding me to be conscious of his own peril. It was an accident much to be regretted, for our isolation was so complete that any loss, however trifling, seemed irreparable by reason of our remoteness from supplies.[5]

Here, then, in microcosmic form, was the fundamental problem with fire power. The same form of energy that could build a city could destroy it; the heat that sustained life could, in excess, snuff it out. "Water and fire are the two mighty elements . . . that form, in some shape or other . . . half our woes and half our felicities," intoned a mid-century copywriter.[6] A popular proverb put it more simply: "Fire is a good servant, but a bad master."

There was nothing new in this dichotomy. Ancient mythology referred to it, as did European folktales. But in nineteenth-century America, where wood was so abundant and the drive for expansion so intense, the juxtaposition formed one of the major undercurrents of contemporary life. In town after town and household after household, work fires repeatedly overstepped their boundaries to consume the productions of their masters. When Boyd referred to "the accustomed smell of burning," she was thinking about her own life. Yet she also spoke for the country at large, where the smell of burning was perpetual and where a vigilant citizenry strove incessantly to distinguish between the sweet scent of progress and the acrid odor of destruction.

Flames raging out of control were a fact of life for a society dependent
on fire for energy. This postcard view from 1908 is typical of
photographs taken across the country.

PROMETHEUS'S PERPLEXING GIFT

The good servant/bad master dichotomy is not the only paradox related to
fire usage. According to the nineteenth-century scholar Thomas Bulfinch,
Prometheus stole fire from heaven and gave it to mankind "out of pity for
their state."[7] But the Titan might well have felt pity for earth's mortals as
they tried to make sense of this perpetually confusing and contradictory
entity.

Consider, for instance, the fundamental incongruity of fire tending—
that fires, though maddeningly difficult to ignite, could also burn with such
ferocity as to be almost impossible to extinguish. Fire users learned this
exasperating truth at an early age, but familiarity with the concept did not
necessarily devalue it through the nineteenth century. References to problems of fire starting and fire stopping crop up in both public and private
writings. In *Roughing It*, for example, Mark Twain satirizes the agonizing
process of coaxing a reluctant fire to life, ruthlessly exposing the shortcomings of such ignition devices as pistols ("an art requiring practice and
experience, and the middle of a desert at midnight in a snowstorm was not
a good place or time for the acquiring of the accomplishment"), rubbing
two sticks together ("at the end of half an hour we were thoroughly chilled,
and so were the sticks"), and, of course, matches, which, though "lovable

9

and precious, and sacredly beautiful to the eye," are in this case entirely ineffective. But starting a fire is not the only problem to plague the hapless narrator of Twain's tale. The author also introduces the flip side of the question when, without warning, a campfire suddenly goes out of control and creates a "roaring . . . popping . . . crackling" scene of destruction. The dryness of the ground cover discourages any attempts at suppression, and four hours later the campers are still watching the voracious flames devour the landscape.[8]

Four hours was a long time, but as newspaper articles and other publications hastened to announce, fires could burn much, much longer. "Burning Yet" was the eye-catching headline for a *Scientific American* account of an urban fire that was still smoldering five months after ignition.[9] Fires that took hold in coal mines or oil wells, where the fuel supply was virtually inexhaustible, could last even longer. In some cases, they smoldered so long that they imparted new meaning to the concept of "perpetual fire." Joseph Husband, a Harvard graduate who spent a year in a midwestern coal mine, published a remarkable account of just such a fire in 1911. Despite the heroic efforts of the miners, it took well over a year to extinguish the fire completely. It is a dramatic story in itself, but some of the impact surely derives from the fact that Husband's readers knew from experience that fires did not always burn so well or so effortlessly.[10]

In truth, fires did not *always* do much of anything. Loge, the Nordic god of fire is generally depicted as a sly, capricious fellow, and the characterization is perfectly apt. A flame could take many forms and produce a wide range of colors, temperatures, odors, sounds, and intensities. More to the point, it did so regularly and "openly"—in full view of society. The result is a profusion of contradictory fire lore that permeates many aspects of American culture and life. It was not uncommon, for instance, for a single word to be applied to two opposing concepts. Thus, *firebag* referred to a bag containing kindling implements as well as to a bag used to save property from a blaze. Similarly, a *fireman* could be either a fire maker or a firefighter, and *fire stick* could denote either a fuel poker or the fuel itself. Surprisingly, aphorisms ran much the same way. A goat was said to "look like a lady" in candlelight, and yet it was claimed simultaneously that a young girl sitting by a fire or stove risked being transformed into a withered old woman.

A broad interpretation of fire artifacts yields additional perplexities. The chimney, for example, functioned primarily as a vent for the smoke and gases emitted from a hearth fire; but sometimes its connotations were more complex. When smoking gently, it was a symbol of prosperity and comfort; when smoking too much, it became a menacing threat to public health and safety. It was the conduit for the beloved St. Nicholas; yet it was also

Panorama from the new Post Office Building, Boston, following the fire of November 9 and 10, 1872. An isolated chimney is a prominent feature of this stereograph.

the mysterious portal for witches, warlocks, and other evil spirits. When built of wood and mud as was common in the early days, a chimney was an ephemeral thing. Its days were clearly numbered, and some communities required homeowners to acquire a long pole to hasten the demise of the flimsy structure in case of fire. Later, with the adoption of brick and stone construction, chimneys became the survivors—often standing blackened and alone to show, as one contemporary put it, that men had "trod this road before us."[11]

Contradictions and complications permeated the world of the firefighter as well. Most intriguing is the fundamental irony that firemen frequently used fire to fight fire. This was true not only in rural areas, where backfires were a favored deterrent to prairie fires, but also in urban centers, where by the second half of the nineteenth century fire-powered steam engines were increasingly used to pump water. Even in their preparations, firemen often had to nurture the flame before quenching it. They set fires in flambeaux to light their way on darkened streets and placed glowing candles in their windows to signal comrades that they had heard the alarm and were on the way. With the advent of steam power and the need for heated firehouses to prevent the boilers from freezing, many firefighters had to set and tend fires on a regular basis. During the antebellum years, some in the fraternity

THE LIFE OF A FIREMAN.

Firemen employed fire, in the form of a smoking, sparking steam boiler, to fight fire. Currier and Ives captured this irony in their lithograph "The Life of a Fireman."

carried this trend to the point of absurdity: some misguided companies of volunteers set fires deliberately, just so they could have the pleasure of putting them out.

Opposites: Recognition and Reconciliation

Hot and cold. Light and dark. Clear and smoky.

Opposites such as these have often been used to describe fire practices— sometimes quotably so, as when Benjamin Franklin observed of fireplace heating that it caused one to be "scorch'd before, while he's froze behind"[12] or when some nameless wit coined the expression "Crooked logs make straight fires."

To Aristotle, such pairings would have made perfect sense. He believed, as a matter of scientific principle, that the recognition of opposites in the natural world is an important first step in reconciliation and understanding. American fire users, by contrast, had no such ambitions. Most of them lacked the intellectual underpinnings and scholarly mindset for such an esoteric approach. But they did have wide-ranging experience with fire. They knew from observation and manipulation that fire is complex, multidimensional, and contradictory. Furthermore, as almost any journal or autobiography of the nineteenth century will show, they seemed to understand that if they were going to use fire power at all, they would have to become reconciled to using it on many different levels simultaneously.

Lucy Larcom personified this accommodation as well as anyone. A worker in one of New England's early mills, Larcom published a memoir that includes a kaleidoscopic array of fire references. She cooked at the fireside and received inspiration in its ruddy glow. Although the lack of a fire "at five o'clock on a zero morning" chilled her physically, she writes that the closing up of the open fire by the introduction of stoves chilled her spiritually. On one occasion, Larcom narrowly escaped injury from some hot coals; yet she continued to enjoy sitting before the open flame.[13]

Boys of the period experienced fire differently than girls, but Edward Everett Hale's childhood was no less affected by fire paradoxes. His autobiography refers to playing with fire and fighting fire, to being burned by fire and being entranced by it. Lincoln Steffens chronicles a similar inventory of experiences: opening Christmas stockings at the fireplace, chasing firemen to fires, and accompanying the "bridge-tender," the man who protected the wooden railroad bridge by extinguishing the live coals that fell from trains. Steffens also describes a confusing paradox that he encountered on a visit to a blacksmith's shop. The sparks that flew off the smithy's hammer spattered everywhere, burning big black spots on the floor but never harming the smithy's arms. How could this be? the boy asked. "They know me," the blacksmith said simply. And then, the dumbfounded Steffens recalls, the man "went on beating the red-hot irons, ducking them sizzling into water and poking them back in the open fire, just as if I wasn't there."[14]

Steffens *was* there, however. And although he recorded the incident for his own reasons, he was also experiencing what seems, in retrospect, to be the greatest fire paradox of all: fires, though separately maintained, were a potent unifying feature of society. Mrs. Boyd, the army wife, suggests the premise of this paradox when she speaks of her isolation and, in so doing, emphasizes one of the many attractions of fire power: independence. As popular wisdom had it, a person could build a fire almost anywhere and survive. Even in populated areas, the notion of the individual household fire was fundamental—so much so that periodic campaigns for cooperative housekeeping were derailed almost before they began. As late as 1913, one woman went so far as to define women's liberation in terms of homesteading and firebuilding. "I am only thinking of the troops of tired, worried women," she wrote, "sometimes even cold and hungry, scared to death of losing their places to work, who could have plenty to eat, who could have good fires by gathering wood, and comfortable homes of their own, if they but had the courage and determination to get them."[15]

An illustration from "Kindling a Fire" captures the tradition of borrowing coals before matches. From Gertrude L. Stone and M. Grace Fickett, *Days and Deeds a Hundred Years Ago* (1906).

On the other hand, although wood and coal fires were inherently discrete, fire as a phenomenon was a great unifier. On the simplest level, the connection among people was merely social. "Did you come for fire?" was a common Pennsylvania greeting,[16] and "May I borrow a light?" has for centuries offered a successful opening for flirtation. But fire also stimulated deeper, more meaningful bonds based on shared experience. Fire maintenance and control probably occupied more individuals than any other type of activity except those related to basic biological needs. Although fire is traditionally portrayed as the symbol of life, the suitability of the metaphor goes far beyond the classic juxtaposition of life with death. Fire, particularly in the context of nineteenth-century American society, can be seen as representing life in all its richness and variety. Fires could be used for good or for evil. They could empower or overpower. They represented work and play, progress and tradition, loss and gain. And they represented these things for everyone.

The significance of this unifying phenomenon is underscored by a simple comparison with electricity, fire's primary successor as an energy source. Whereas the power lines of today clearly connect individuals in a physical sense, a notion of interconnectedness among the populace is

This charming domestic scene, entitled *The Volunteer Fireman*, was painted ca. 1840 by William Sydney Mount and aptly conveys the two sides of fire. The fireplace, on the left, is balanced by an uncontrolled blaze outside the window to the right. Courtesy of The Museums at Stony Brook, New York; bequest of Mr. Ward Melville, 1977.

clearly lacking. Indeed, some would argue that one of the hallmarks of modern American life is alienation of the individual from government, from neighbors, even from members of one's own household.

Henry Adams had a vague premonition that profound changes were coming to American as the twentieth century began; and although his ruminations are wide-ranging, he knew that energy production had to figure in the analysis. Attending the St. Louis Exposition in 1904, Adams took careful note of the dazzling electrical display. "The world had never witnessed so marvellous a phantasm," he wrote; "by night Arabia's crimson sands had never returned a glow half so astonishing, as one wandered among long lines of white palaces, exquisitely lighted by thousands on thousands of electric candles, soft, rich, shadowy, palpable in their sensuous depths."[17] Adams was unsure what this opulent display meant. Did the "extravagance," as he termed it, reflect the past or did it "image" the future? Was it a creation of the old America or the new? Adams did not know. What he did know, however—from a lifetime of observation and study—

was the essence of the "old" America. And, as Adams's autobiography makes abundantly clear, one of the distinctive features of the country during its formative years was its dependence on fire.

Adams speaks fondly of his own eighteenth-century fire heritage: of his grandfather's "flint-and-steel" on the mantelpiece, of the fact that this illustrious grandfather used to light his own fires, and of the telling point that the "eighteenth century," in this cultural sense, lasted well into the nineteenth. He speaks of changes too—of the improvements in heating and lighting that were introduced into his Boston house and of the curious fact that, despite the obvious technological superiority of the urban dwelling, he "liked it no better for that." As far as industrial fires are concerned, Adams credits them with nothing less than the transformation of the American landscape. As he saw it, steam power surged across the countryside, wresting acreage from farmers and substituting a gritty world in which "tall chimneys reeked smoke on every horizon, and dirty suburbs filled with scrap-iron, scrap-paper and cinders, formed the setting of every town."[18]

This book is about fire in America in the fifteen or so decades that Adams claimed as his own. From the founding of the Republic until well into the twentieth century, most people were keepers of the flame. Many of the

CENTENNIAN COOK.

This cook stove trade card acknowledged the central role of the stove in American family life. Although the technology changed between 1776 and 1876, the fire remained a constant.

technologies related to fire management and control changed during these years—most notably in terms of the gradual, but steady, insulation of the user from the ferocity of the open flame. Nevertheless, the power of fire remained strong, and the power of the *idea* of fire stronger still.

The details of this story are many, ranging from tools and fuels to domestic science and urban planning. But the complexity of the particulars merely underscores the simplicity of the basic, underlying truth: fire has been a profoundly important influence in American society, and Americans themselves, though segregated in many ways, have been inextricably linked by their fire habits, by their fire losses, and, most of all, by the supreme confidence that they could harness the power of combustion and subject it to their will.

Good Servant

> Long, long ago, when people first lived on this earth, they did not
> know about fire. They had no fire to keep them warm. They had no
> fire to cook their food. If they ate fish or the flesh of animals, they ate
> it raw. They could not roast or boil or fry it. You would not like that.
> After a time someone discovered a way to make fire. Perhaps it was by
> rubbing two sticks together. . . . Now we think we could not live with-
> out matches and fires.[1]

This tantalizingly brief history of fire appeared on the back of a children's
stereograph from about 1900. The story seems almost comically spare, especially in light of recent scientific research. Archaeologists now believe,
for instance, that humanoids began to control the energy of fire more than
a million years ago, after they had already learned to use naturally occurring fires, such as lightning-induced and volcanic flames.[2]

Nevertheless, the card was essentially correct about two things. At some
point during the Paleolithic period humans discovered a way to make fire
(first, it now appears, by striking flint against pyrite to obtain sparks but
eventually by "rubbing two sticks together" in the classic friction style).
And, as the card implies, this development was of monumental significance; in fact, many modern researchers consider it to be the most momentous occurrence in human prehistory.

We can easily see why. Fire is produced when sufficient heat is applied
to a fuel in an atmosphere containing oxygen. Human beings eventually
learned that this chemical reaction not only can be perpetuated easily but
also can be exploited in countless productive ways. Heat and light are the
most common by-products of combustion, but smoke, ashes, and air currents also result and can be put to use.

In the largest sense, of course, the use of fire for food preparation or
metal smelting is fundamentally the same whether the flames are stoked in
Asia or the Americas, in the fifth century B.C. or in the nineteenth century
A.D. But as Sir James George Frazer, Walter Hough, and other scholars
have shown, an understanding of different cultures' overall fire-use patterns can be enormously instructive.[3] This is as true for American society

as for any other, and an exploration of the nation's fire habits from the establishment of the Republic to the widespread introduction of electric power offers insights into work patterns, domestic arrangements, and overarching social values. An overview of the extraordinary range of fire-related activities in the nineteenth century provides a good starting point for such an exploration. The purpose of the following examples is not to illuminate fire technologies themselves but to highlight their ubiquity. In homes and workplaces across the country, fire supplied the energy that got the job done. And, as this survey reveals, many separate fires were often required by individual workers to accomplish their goals.

FIRE POWER

Home Fires

Nineteenth-century poets often depicted the hearth as the heart of the home, a logical designation since the words are related.[4] As every housewife knew, however, the kitchen fireplace was also a seething power plant and the household's bustling workplace. Domestic fires roasted the meat, boiled the soup, and baked the bread. They heated the water needed for washing, scalding food containers, and making glue. Their radiant energy dried everything from clothing and feathers to tinder, gunpowder, firewood, and hair. Though open fires were woefully inefficient as area heating devices (modern studies show that only 12 to 14 percent of the heat generated gets into a room), they warmed at least the front half of those sitting nearby. Additionally, they provided the hot coals used in warming pans, smoothing irons, and foot warmers, and, until well into the nineteenth century, they were the primary source of sparks needed to light pipes, ovens, and candles. Cast-iron stoves took over many of these functions toward the middle of the century, but it was still fire—now enclosed in metal—that heated rooms, cooked food, warmed irons, and evaporated water to humidify the house.

Whether glowing in the fireplace or dancing freely at the end of a lamp wick or gas jet, fire illuminated the home. It also melted sealing wax for letters, heated curling irons, and facilitated the separation of honey from the comb. Wood ashes were used to scrub floors as well as to sterilize glassware, fertilize plants, prepare lye, and create that staple of pioneer fare, hominy. Popular household manuals suggested a myriad of additional uses for fire and its by-products. "A hot shovel or warming-pan of coals held over varnished furniture will take out white spots," advised one such treatise. "Glass vessels in a cylindrical form, may be cut in two, by tying around them a worsted thread, thoroughly wet with spirits of turpentine, and then

The fireside was a place for families to gather and read. The manufacturers of "The Fireside" furnace took advantage of this American tradition in their advertising card, ca. 1880.

Wood- and/or coal-burning stoves made their way into households across America during the nineteenth century. Colorful trade cards and fanciful names were common advertising ploys.

setting fire to the thread," claimed another. Lydia Maria Child's popular guide, *The American Frugal Housewife* (1833), described how one woman even used fire to rid her home of an army of red ants. After luring the ants to a dish filled with shagbarks, she simply dumped the ants, bait and all, into the fire. (Several treatments were required.)[5]

Reading by candlelight was a way of life for many Americans, although this image, entitled "Ghost Story," graced an advertising trade card of the 1870s, long after candles had given way to other forms of lighting.

In the absence of refrigeration, fire was essential for food preservation. Heat dried herbs and vegetables, and the judicious application of smoke and ashes allowed meat to be kept for prolonged periods. Smoking could be accomplished inside a chimney or in a separate smokehouse. The procedure varied according to the type of meat and its size. For legs of mutton, Lydia Child advised: "Lay them in the oven, on crossed sticks, and make a fire at the entrance. Cobs, walnut-bark, or walnut-chips are the best to use for smoking, on account of the sweet taste they give the meat. The smallest pieces should be smoked forty-eight hours, and large legs four or five days. . . . Meat should be turned over once or twice during the process of smoking." Hams required additional treatment—as many as four weeks of smoking, preceded by liberal salting and followed by careful packing in barrels filled with ashes or charcoal. Smoked meat could also be hung in the cool darkness of summertime chimneys.[6]

As these procedures imply, American householders were active producers of many commodities, and fire—on the hearth or in rudely constructed outdoor facilities—was central to many such endeavors. It provided the heat required for the manufacture of dye and cheese; for hard-working New Englanders in particular, it facilitated the transformation of the family fireside into a primitive forge for the hand finishing of nails during the winter months. Although it was possible to make "cold" soap without the use of "artificial" heat, housewives everywhere participated in the annual ritual of combining lye (which they had made themselves using their

New forms of lighting fuel, introduced during the nineteenth century, created new styles of home lighting.
Illustration from A. D. Jones, *The Illustrated American
Biography* (1853). National Museum of American History, Smithsonian Institution Photo 38915.

own wood ashes) with animal grease over a roaring open fire in order to
produce their family's yearly supply of this essential product. Fire also
helped to make candles, although the tedium of this process—one pioneer
claimed that during the 1820s it took three or four hours to hand-dip six
candles—induced many families to opt for commercially produced tapers
as soon as they could afford them.[7] Sugar making, on the other hand, was
one fire-related task that was nurtured generation after generation and
then survived even longer in the nostalgic recollections of the participants. "What days and what nights were spent by us in that sugar house or

camp, as we called it, boiling down sap," recalled John E. Brackett of his Vermont childhood in the 1850s. "At night, [we sat] or stretched out at full length on the ground around the fire listening to Uncle Bill's stories of Indians and of wars." Uncle Bill, meanwhile, monitored the kettles and periodically allowed the children to sample the sweet syrup.[8]

Though not so fondly remembered, home remedies, too, called for manipulation of the flame. Fire cooked up a plethora of medicines and ointments and also distilled the alcohol that figured prominently in many of these concoctions. It heated the irons used to cauterize wounds and, in the case of a more sophisticated implement called a moxe, was ignited within the device itself so that therapeutic burning could continue as long as necessary.[9] Although the usefulness of fire as a disinfectant was inadequately understood throughout much of the period under discussion, "smoak-houses" were occasionally erected in the late eighteenth and early nineteenth centuries to cleanse victims of smallpox. Later, in response to the diphtheria epidemics of the 1890s, local authorities frequently ordered that the clothing and even the houses of infected persons be burned in order to halt the spread of disease.

Folk medicine picked up where conventional medicine left off and promoted a multitude of fire-based therapies. A spoonful of ashes stirred in cider was thought by some to prevent stomachaches as well as cholera in the 1830s. Smoke—inhaled, swallowed, or blown in the ear, depending on the ailment—was widely proclaimed as a cure for goiter, headaches, earaches, and colic. "Burn orange peelings on the stove to cure or to prevent influenza," advised one Nebraska pioneer. "Heat a seeded raisin very hot over a kerosene lamp or a fire . . . [and] put the raisin into [an] aching ear," counseled another. Moving almost imperceptibly from the world of medicine to the world of superstition, lay physicians also recommended the following:

FOR A TOOTHACHE

Cut the first skin off the frog of a horse's front hoof and put the skin over a fire until it chars. Crumble it and put it on the aching tooth.

FOR A MAN'S HEADACHE

Put a bowl on his head. Then cut his hair around the bowl, and burn all the hair. This will stop headache.

FOR A NAIL IN THE FOOT

Pull the nail out, grease it, and throw it in the fire. This will keep the foot from getting sore.[10]

Living Off the Land

The earth offered a good living to those who knew how to handle fire. The use of the smudge for hunting, for example, has many variants, one of the more creative of which appeared in the South during the last century. As described by ethnologist Walter Hough, these enterprising hunters placed a burning cotton wick on the back of a turtle and then waited until the turtle had wandered far enough into the underbrush to smoke out a woodchuck.[11] More dramatic, perhaps, was the true "fire-hunt" in which firelight was used to stun an animal before the kill. As Elizabeth Ellet described the process in 1852, a horseman rode through the woods carrying a "fire pan" full of blazing pine knots, which cast "a bright and flickering glare far through the forest." A second man followed and, as soon as a deer or other large animal had become mesmerized by the brilliance, took aim and fired.[12] Hunters with less finesse or a disinclination for night work simply set the woods on fire and shot their prey during the ensuing stampede.

As Richard Henry Dana pointed out in his autobiography, *Two Years Before the Mast* (1835), it was also possible to fish with fire:

> For this purpose, we procured a pair of *grains*, with a long staff like a harpoon, and making torches with tarred rope twisted round a long pine stick, took the only boat on the beach, a small skiff, and with a torch-bearer in the bow, a steersman in the stern, and one man on each side with the grains, went off, on dark nights, to burn the water. This is fine sport. Keeping within a few rods of the shore, where the water is not more than three or four feet deep, with a clear sandy bottom, the torches light everything up so that one could almost have seen a pin among the grains of sand. The craw-fish are an easy prey, and we used soon to get a load of them. The other fish were more difficult to catch, yet we frequently speared a number of them, of various kinds and sizes.[13]

"Burning the water" was primarily an amusement, but for most American farmers burning the land was serious business. Prescribed burning of forests and grasslands was widely practiced by Indians and Europeans alike in their efforts to clear land, increase soil fertility, and reduce fire hazards in rural areas.[14] "Father would chop and grub all day, then he would burn brush and log heaps until ten or twelve o'clock at night," wrote one farmer's son, recalling the slash-and-burn technique that remained almost as common in the Midwest of the early nineteenth century as it had been centuries earlier in New England.[15] That this approach was wasteful, there is no doubt. "I helped to burn, and make into rails, timber that would now

LIFE IN THE WOODS.

These hunters may not have used fire to lure their prey, but they used fire
nonetheless. Peters Collection, National Museum of American
History, Smithsonian Institution Photo 60460-A.

be worth more than the land," one Midwesterner recalled with dismay at
the end of the nineteenth century.[16] And yet, because fire could clear a
field so much more quickly than an axe alone (an Indian claimed that an
industrious woman could burn in a day as many logs as a man could chop
in two or three),[17] combustion remained a valued aid to soil preparation.
After that, fire and its by-products facilitated numerous agricultural endeavors: candles illuminated barns, ashes killed pests on crops, and smoke
applied to tobacco leaves rendered them ready for market.

Fire also figured prominently in various land management policies outside the world of the farmer. Broadcast burning, for instance, was advocated and practiced for years despite society's obvious inability to control
wildland fires with precision.[18] It was far safer, of course, to dispose of small
amounts of rubbish in carefully monitored fires, and well into the twentieth century this was the standard method of disposing of debris for private
individuals and communities alike. Less frequently but no less effectively,
fire was used to clear streets of ice and snow and even, on occasion, to
destroy unwanted buildings. (Some enterprising seventeenth-century Vir-

ginians used fire to burn down unoccupied frame houses in order to recover their precious nails, but this practice was quickly outlawed.)

Interestingly, fire was not used extensively in early America to dispose of the dead. Although cremation dates back to prehistoric times and found favor in a variety of cultures thereafter, it ceased to be common in Europe during the Christian era owing to the strong emphasis on the doctrine of the resurrection of the physical body. Sporadic instances of cremation occurred in the European and Indian sectors of American society from the seventeenth century on, but only in the last quarter of the nineteenth century did the practice begin to attract serious interest in an organized fashion. F. J. Le Moyne built a small furnace for funerary purposes in Washington, Pennsylvania, in 1876; eight years later, in 1884 (the same year that cremations were legalized in Great Britain), America's first crematorium was constructed in Lancaster, Pennsylvania.

Industrial Fires

Meanwhile, augmenting such small-scale fires, industrial-strength combustion animated virtually every sector of the world at large. Deep within the earth, fires illuminated the nation's mines, energized the associated subterranean blacksmith shops, and, if there were enough candles left over for luxuries, heated the miners' coffee in chafing-dish style.[19] Though fraught with danger, fire also figured in various schemes for mine and tunnel ventilation. As engineers had discovered, a roaring fire sustained at the foot of an air shaft was all that was needed to create the life-sustaining air currents required for prolonged underground work. The blaze created hot air and gasses that, as they ascended through the chimney-like shaft, caused the stagnant air to circulate below.[20]

Ignited as easily above the ground as below, fires glowed at the top of lighthouse towers to warn ships away from dangerous coastlines and burned at the top of shot towers to melt lead ore prior to its sphere-producing descent to the ground. Occasionally, fire even defied gravity. Airborne fires not only created exotic pyrotechnic displays but also helped to launch many of the earliest hot-air balloons that excited Europeans and Americans toward the end of the eighteenth century and the beginning of the nineteenth.[21]

It was on firm ground, however, that commercial fires were most prominent, powerful, and even most recognizable, since in many instances commercial fires were simply more sophisticated versions of the homemade variety. "What is accomplished by fire is alchemy, whether in the furnace or kitchen stove," the sixteenth-century Swiss alchemist Paracelsus is said to have remarked,[22] and, indeed, fire-generated heat could just as effectively smelt metal as bake bread. Sometimes, in fact, there was very little difference between the two processes. Just as bread was produced on the hearth by placing dough in a Dutch oven and covering the pot with hot ashes, so lead was easily retrieved in Missouri during the 1820s and 1830s by placing the ore in a three-sided fireplace (minus the chimney) and covering it with burning logs. (That these log heaps extracted the metal "at a great waste of fuel," as one mining consultant wrote in 1842, merely underscores the comparison with hearth bread.[23]) Other commodities required higher and more evenly controlled temperatures and, hence, more concentrated fuels and more sophisticated furnaces. But whatever the apparatus, fire was the powerful agent by which one kind of material could be magically transformed into another.

Not surprisingly, the number of materials that underwent transformation during the last century is considerable. "Nearly all operations carried on in the arts require the application of artificial heat," wrote the world-renowned chemist Friedrich Knapp in his 1848 treatise that described state-of-the-art procedures for manipulating fire in the production of pottery, glass, brick, salt, boric acid, and quicklime.[24] Fire also figured prominently in the commercial manufacture of dye, sugar, candles, and potash and in the distillation of perfumes, turpentine, and any number of beverages. Still other applications included the processing of agricultural products, for which enterprising manufacturers sold a variety of heavy-duty "agricultural boilers." Dentistry and jewelry making were also aided by

"AGRICULTURAL FURNACE & BOILER."

Specially designed boilers were offered to "butchers, farmers, cheesemakers, stock raisers, hotel keepers, bakers, brewers, chemists, druggists, dyers, painters, laundries, chandlers, and for various other manufacturing and mechanical purposes." Trade card by J. S. and M. Peckham, Utica, New York, ca. 1865.

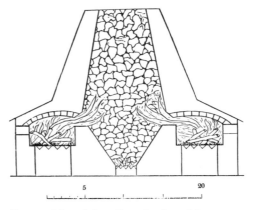

Perpetual lime kilns allowed calcined limestone
to be removed and new limestone added without
cooling. Designs for perpetual kilns varied, but
most had several ovens around the perimeter of
the chimney. From M. Tuomey, *Report on the
Geology of South Carolina* (1848).

Chain Shop, Mass. State Prison.

Furnaces were adapted to many specialized tasks. This Massachusetts State Prison shop,
illustrated in a stereograph ca. 1880, produced chains to restrain the prisoners.

specially designed "melting furnaces." Although virtually all metal workers relied on fire for assaying, smelting, and finishing their products, the intimate connection between fire and iron was so entrenched in the public mind that it was widely celebrated. Artists depicted blacksmiths laboring over ruddy coals, and poets glorified both the ironmaster's "fires in the wilderness" and the "flaming forge" of the artisan who hammered the metal into shape.[25]

In addition to generating heat, fire provided light for America's industrial and commercial enterprises; and as in the case of heating, many of these lighting devices were clear adaptations of domestic models. "Weaver's lamps," for instance, were often no more elaborate than ordinary candles mounted on special holders that could be hung on a loom. Likewise, the "candling" of chicken eggs was accomplished simply by holding an egg in front of a candle or small lamp in a dark room. Despite the gradual introduction of powerful lenses in lighthouses, many beacons at mid-century were basically oil lamps, albeit with numerous and exceptionally wide wicks.

On the other hand, special situations called for special illuminants. Stage technicians, for instance, borrowed the now-famous limelight from the world of land surveyors as early as the second quarter of the nineteenth century. Created by heating sticks of pure lime to high temperatures by means of an oxygen-hydrogen flame, this intense light proved to be as effective for illuminating a darkened stage as it had been for signaling across great distances outdoors. So brilliant was the effect, in fact, that limelight remained a popular form of theatrical lighting long after the advent of electricity.[26]

The miner's safety lamp is another example of lighting adapted to particular occupational needs. Customarily, miners illuminated their work areas by affixing candles to their helmets with clay. Such measures seemed perfectly satisfactory until underground coal mining operations demonstrated a deadly scientific fact: an open flame in contact with the coal-generated gas methane (miners called it "firedamp") produces a highly explosive situation. Alarmed by the mounting casualties from this cause, the British scientist Sir Humphry Davy designed a lamp that circumvented the problem by shielding the flame from the dangerous gas by means of a metallic gauze. The open-weave design successfully prevented the air surrounding the flame from reaching the high temperature required for explosion. The lamp enjoyed widespread use in American mines throughout the nineteenth century.

Early experiments with gaslight conducted by Rubens and Rembrandt Peale in their museums furnish yet another example of the adaptation of new lighting technologies to particular occupational needs. The usefulness of carbureted hydrogen gas as a fuel for street lighting had been demon-

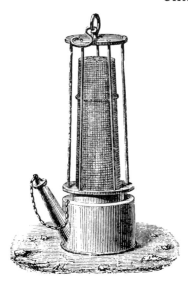

Sir Humphry Davy's miner's safety lamp reduced the risk of coal mine explosions by surrounding the lamp flame with a wire gauze that reduced temperatures of potentially explosive gas below the flash point. Illustration from A. Cazin, *The Phenomena and Laws of Heat* (1869).

strated in London as early as 1807. Almost immediately, Rubens Peale determined that such a system would not only enhance the scientific reputation of his Philadelphia exhibition hall but also present an irresistible attraction for the paying public. It took nine years and $5,000 to accomplish his goal, but in April 1816 Peale treated an enthusiastic nighttime crowd to a brilliant display of gaslight issuing forth from five huge burners and augmented by an array of glittering cut glass. A year later, Rembrandt Peale introduced a similar system to his Baltimore museum. In this case, the main gallery boasted a "magic ring" of fire that consisted of a hundred individual flames that expanded or contracted by the simple rotation of the fuel valve. The public flocked to these displays and so helped defray the cost of installation.[27]

Gas lighting quickly spread beyond the confines of these museums to the world at large—to street lighting in Baltimore as early as 1817 and eventually to public buildings and private homes in urban areas across the country. There was, however, one large-scale industrial use of fire that almost completely circumvented home applications and yet profoundly influenced the industrial and transportation systems of the entire nation. Some early writers referred to the mechanism as a "fire machine," but the more common appellation quickly became "steam engine."

The first practical steam engine was constructed toward the end of the seventeenth century by Thomas Savery for the purpose of draining water from mines. Although this prototype did not accomplish the stated goal

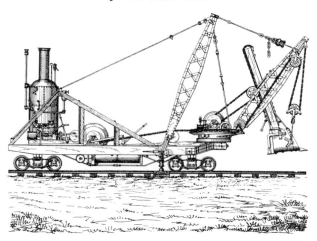

Steam power found dozens of industrial uses. Benjamin
Butterworth illustrated this 1884 patent steam excavator in
his compendium, *The Growth of Industrial Art* (1892).

with any efficiency, successive models, which benefited greatly from the
innovations of Thomas Newcomen and James Watt, were gradually introduced into mining operations in Europe and, to a lesser extent, America
during the eighteenth century. From this subterranean debut, the steam
engine went on to conquer the earth's surface, where by the second half of
the nineteenth century it had virtually revolutionized American society.
Fueled initially by wood and later by coal, steam engines provided the
power for the nation's railroads, steamboats, and, eventually, warships.
Steam drove fire engines, plows, mowers, reapers, shovels, and road rollers.
In addition to such mobile activities, stationary steam engines accomplished tasks as diverse as pile driving, milling, and hoisting. The Centennial Exhibition of 1876 promoted George Corliss's monumental fourteen-
hundred-horsepower steam engine as one of its major showpieces, but this
was merely the most flamboyant example of the ubiquitous machine that
lent its name to an epoch.

Assessing American accomplishments at mid-century, a patriotic con-
tributor to *Scientific American* judged steamships and steam locomotives to
be two of the three greatest feats of "practical science" of the day (the third
was lighting by gas). The writer pointed out that each of these achievements had been denounced as "utterly impracticable" by the nay-saying
philosophers of the day. What he did not bother to mention—possibly
because it was so obvious—was that each of these remarkable accomplishments relied on fire.[28]

Steam tractors, which provided farmers with a means to cultivate more acreage, were in widespread use by the end of the nineteenth century. Stereograph by Underwood and Underwood, 1904.

An 1879 patented steam harvester combined animal power for pulling with steam power for harvesting the grain. From Benjamin Butterworth, *The Growth of Industrial Art* (1892).

The Fires of Genius

American miners and quarry workers of the last century occasionally used fire as a means of splitting rocks into more manageable pieces. The builders of King's Chapel in Boston, for instance, built small fires directly on top of large stones and then, when the surfaces were deemed hot enough, dropped iron balls onto them to effect the cut.[29]

Scientists also cut rocks with fire, but their purpose was more inquisitive. Like latter-day alchemists, they wanted to see what was inside. It seems fitting to mention, therefore, that in addition to its centrality to ordinary life, fire was also an indispensable tool in the rarefied atmosphere of the scientific laboratory. In fact, a laboratory was not even necessary. Charles Goodyear made his important discovery concerning the vulcanization of rubber after observing the effect of heat on a piece of sulfur-treated rubber that accidentally had been allowed to touch a hot stove. More formally, organized research centers of the nineteenth century provided any number of specialized fire-making devices that greatly facilitated inquiries into the nature of matter and energy. The Bunsen burner is probably the most famous such apparatus. Named for (but not invented by) the German chemist Robert Wilhelm Bunsen, the Bunsen burner combines air and gas in such a way as to produce a hot but nonluminous flame. Other fire-making devices of the last century include a variety of blowpipes, burning glasses, and small furnaces—all exploited to the full by a cadre of investigators driven to unlock the secrets of their universe.

THE POWER OF FIRE

In pre-electric American society the ubiquity and potency of fire endowed the flame with a significance that far surpassed the utilitarian. Archaeologists, in fact, might argue that fire enjoyed two levels of function beyond the purely technical: first, a "sociotechnic" usage, which places fire in its social setting; and, second, "ideotechnic" applications, which encompass fire used or referred to in religious or ideological contexts.[30] Examples of both categories of fire flicker again and again in the American historical record. Indeed, even disregarding religious functions entirely, it is clear that fire was an extraordinarily versatile servant. Even as it was employed to smelt metal and warm the home, it was also exploited for fun and for profit. To grab attention, to amuse, to inspire fear, and simply to inspire— these must be counted among the manifold uses found for the flame since the earliest days of settlement. Americans have been celebrating fire as long as they have been harnessing its power.

Playing with Fire

THE LURE OF THE LURID GLOW

For sheer excitement, few diversions could match the drama of a real conflagration. That was Mark Twain's conclusion, at any rate, when his campfire suddenly flared out of control and created a scene of destruction that
was strangely compelling. "Within half an hour all before us was a tossing,
blinding tempest of flame!" he wrote. Fascinated, he and his companion
promptly retreated to a boat and watched the spectacle.[31]

Equally dramatic were the annual prairie fires that swept across the dry
grasses of the Midwest each autumn. "Words cannot express the faintest
idea of the splendor and grandeur of such a conflagration at night," recalled one Indiana pioneer in 1881. But language served him well as he
conjured up a pale queen of night who "dispatched myriads upon myriads
of messengers to light their torches at the altar of the setting sun until all
had flashed into one long and continuous blaze."[32] To another settler,
theater imagery came naturally to mind when discussing prairie fires: "The
gentle breeze increased to stronger currents, and soon fanned the small,
flickering blaze into fierce torrent flames, which curled up and leaped along
in resistless splendor; and like quickly raising the dark curtain from the
luminous stage, the scenes before one were suddenly changed, as if by a
magician's wand, into one boundless amphitheatre, blazing from earth to
heaven and sweeping the horizon round."[33] It is easy to understand why
the Count in Washington Irving's *A Tour on the Prairies* considered a
prairie fire to be a mandatory attraction during his trip and why, despite a
firm commitment to fire control, citizens across the country roused themselves when they heard the fire alarm—not so much to help save property
from the flames, but rather, as Henry David Thoreau put it, "to see it
burn."[34]

This army of onlookers could be found in small towns and villages
everywhere, but it was especially prominent in the larger cities, where the
frequency of alarms gave ample opportunity for fire engine chasing. New
York's "loaferage," as George Templeton Strong called them, were particularly adept at their sport. "They consider [fires] a sort of grand exhibition
(admission gratis) which they have a perfect right to look at from any
point they like and to choose the best seats to see the performance," Strong
remarked in his diary. "The interests of the owners never seem to enter
their heads, and any attempt to keep them back, or to keep a passage open
. . . they consider an unwarrantable interference, of course." Strong was
something of an expert on the subject. This New York lawyer attended
fires regularly, and, like the rest of the loaferage, he looked at a blazing

Currier and Ives's dramatic representation of the burning of New York's
Crystal Palace, October 5, 1858, prominently features the
large crowd that gathered to watch the blaze.

building and saw a melodrama. Strong's descriptions, in fact, read more
like reviews than reports. On February 3, 1840, for example, he witnessed
"a good, steady, old-fashioned conflagration, in which the dramatic interest was well-sustained throughout, and fire and water were 'head on head'
till the grand finale when the walls tumbled down in various directions
with a great crash, and then fire triumphed." Since the fire was the "hero
of the piece," concluded Strong, "it was very proper and perfectly regular
that it should do so."[35]

Proper or not, few spectators would have disagreed that the fire itself was
the main attraction. Even the dry journalistic accounts of the day seem to
labor long and lovingly over fire descriptions, conjuring up the "lurid
glare," "the roaring and crackling of the forked flames," and, in the case of
the San Francisco fire of 1851, the burning metals that "curled up like
scorched leaves, and sent forth their brilliant flames of green, blue and
yellow tints, mingling with the great red tongues of fire which flashed
upwards from a thousand burning buildings."[36]

There were plenty of other reasons to dash to a fire, and the public
thought of most of them. For many, the thrill came not so much from

8233. Burning of the Cold Storage Building, Fifteen brave Firemen lost their lives, July 10th, Columbian Exposition.

Even though fifteen firemen were killed fighting the fire at the Cold Storage Building, most people viewed the destruction as an afternoon's entertainment at the Columbian Exposition in Chicago, 1893. Stereograph by B. W. Kilburn, Littleton, New Hampshire.

viewing the flames as from watching the firemen do battle with them. Basil Hall, an Englishman who visited New York in 1827, claimed a purely academic interest in observing the skills of the city's famous firemen. Thus, when he followed the crowd to a fire in the middle of the night of May 20, he used the opportunity to assess the effectiveness of the firefighting operations, which, he complained, were marred by "needless shouts" and "fool-hardiness." Other observers intellectualized less about their activities. Shouts and foolhardiness were to be expected since, in their view, firefighting was primarily a public show. Many spectators, in fact, congregated simply to cheer on their favorite companies.[37]

Children across the country had their own agendas at fires. A young girl in California used the occasion to flirt with a neighbor, whereas William Dean Howells preferred to watch the feeble spluttering of the ineffective engines or, better still, see the hose break. Edward Everett Hale aimed higher, because in Boston boys were sometimes permitted to help drag the engine or work the brakes. On the other hand, if such pleasures were denied, as they generally were to the smaller boys, there remained the less important but very real satisfaction of attending a conflagration and shouting "Fire!"[38]

Children enjoyed watching fires and firemen, though they had to be careful not to get in the way. Lithograph by Harrison and Weightman, Philadelphia, 1858. Peters Collection, National Museum of American History, Smithsonian Institution Photo 60451-A.

Even the aftermath of a great fire had crowd appeal. Tourists flocked to the scenes of devastation just to see the lone chimneys, mounds of ashes, and holes in the ground. In wintertime, when the firemen's water froze almost as soon as it hit the buildings, there was the added attraction of cascades of icicles hanging from the skeleton-like structures. As P. T. Barnum recalled following the March 1868 fire at his museum, "thousands of persons congregated daily . . . to get a view of the magnificent ruins," particularly the "gorgeous frame-work of transparent ice," which, Barnum conceded, was even more "sublime" by moonlight.[39]

The public was also lured by the remote, but very real, possibility of hearing firsthand accounts of the disaster. Such was the case after a huge fire in the town of Laurel, Delaware, in June 1899. "The visitors wander around and eagerly listen to all tales of personal loss or thrilling experiences related by the sufferers," noted a local newspaper. "When one of the

Spectators gathered after the great Boston fire of 1872.
These ruins were all that was left of once-thriving Pearl Street.

Spectators were fascinated by the lace work of icicles covering P. T. Barnum's
burned out American Museum in 1868. From Barnum, *Struggles and
Triumphs: or, Forty Years' Recollections of P. T. Barnum* (1869).

Selling Relics of the fire

Street vendors capitalized on the Chicago fire of 1871 by setting up impromptu stands and selling relics of the fire. Fused masses of nails, twisted and melted kitchenware, and other oddments of the intense heat were offered for sale. Stereograph courtesy of the Russell Norton Collection.

latter begins a narrative, he is quickly surrounded by a crowd of eager auditors, who listen with deep interest to his story and ply him with all sorts of questions."[40]

Even at the remove of many decades, such gawking has a slightly unsavory aspect to it. The lookers were not as base as the looters, but they did have a disquieting tendency to find amusement in the misfortunes of others. Worse, the crowds constantly disrupted the operations of firefighters and sometimes even put themselves at risk. In Syracuse, New York, thirty bystanders were killed when some gunpowder exploded during a fire.

It could be argued that the urge to be at the center of the action was not just human nature but a predictable outcome of volunteer firefighting systems, in which, in Hale's words, "it was everybody's business to attend at the fire."[41] Certainly, in most small towns and in cities with unpaid departments help was generally welcome at a blaze. Sometimes attendance was even mandated by law. All too often, however, these extra hands— especially if there were several thousand pairs of them—brought confusion rather than assistance. As early as 1715 a Bostonian complained about the "lookers on" at fires and suggested revamping the entire firefighting system to alleviate the problem. New York tried a different solution and tightened its police protection at big blazes. To a resident of Pittsburgh, however, must go the prize for the most ingenious suggestions of all. Writing to the *Pittsburgh Gazette* in 1809, the self-styled "Admirer of a Fire" offered two methods of handling the crowds. One, denoted "plan dilatory," stipulated that the city burn all its firefighting equipment so that the show could continue uninterrupted. The other, "plan Immediate," called for the establishment of a municipal "conflagration fund" to be used to buy twelve houses each year. Then, month by month, the authorities would torch the structures in what was sure to be a glorious celebration.[42]

These were no solutions at all, as the satirist knew. But at least this admirer's plans had the merit of recognizing and accepting human nature for what it was. All that was really needed was for someone to package exhibitions of fire in more convenient, socially acceptable ways. That many such entrepreneurs appeared on the scene is scarcely surprising. The range of their showmanship, however, is extraordinary.

FIERY DISPLAYS

Although New Yorkers had heard music with a fire motif before, they had never heard or seen anything quite like the "Firemen's Quadrille." Antoine Jullien, a noted French conductor, unveiled his contribution to the genre during a concert series presented in New York City in the summer of 1853. The piece began innocently enough. Just a few short measures after the opening, however, the music erupted with strident sounds from the brass section and the clanging of real fire bells. Then, much to the astonishment of the audience, flames shot from the ceiling of the theater and, in prompt response, three companies of firemen charged down the aisles with their ladders and hoses. Scurrying about and shouting, the firemen sprayed water on the flames, shattered false wooden panels with their axes, and splintered glass. Despite reassurances by the ushers that the fire was part of the show, the audience began to panic. Things were clearly getting out of hand, so Jullien directed the orchestra to switch to the *Doxology*, a well-meant selection that only confirmed the crowd's suspicions that the end was near.[43] Order was restored eventually, but thereafter Jullien—and most other bandleaders—preferred to offer "fire music" that featured little more than fire bells and other raucous sound effects.

For the real thing, audiences had to look elsewhere. Fire being the enormous attraction that it was, however, they did not have to look very far. For more than half a century a small cadre of performers had been developing a roster of death-defying stunts that pitted human beings against the fury of the flame.

The earliest tricks of these so-called fire-resisters were fairly rudimentary. During the eighteenth century, for instance, touring exhibitionists ate live coals before an incredulous public, and in St. Louis in 1814 one Leitsendorfer claimed to be able to put a live coal on his foot and then toss it into his mouth without injury.[44] From these modest beginnings, the art form expanded and evolved until performers were routinely stepping on hot coals and sitting in hot ovens.

The French showman Ivan Ivanitz Chabert had become the acknowledged master of these feats by the second quarter of the nineteenth century. As his publicity posters advertised, Chabert could climb into an oven heated to more than five hundred degrees and remain there until the steak he carried with him was cooked to perfection. Other stunts performed by the "Fire King" included holding a lighted candle under his feet for several minutes, eating a torch "as if it were a salad," and ingesting poison.[45] Chabert presented these and other sensational acts to audiences in England during the 1820s. Then, in 1831, he took his "Temple of Fire" to the United States, whereupon he not only attracted enthusiastic crowds but also provoked the appearance of an American challenger named W. C. Houghton. A tireless performer, the "American Fire King" defied Chabert at every opportunity. Traveling up and down the east coast, he boasted endlessly that he could execute "fiery feats" considered by Chabert to be impossible.[46]

Not surprisingly, the only clear winner to emerge from this publicity campaign was the fireproof act itself. The famous American magician Richard Potter perfected it—sometimes running his hands through flames, sometimes bending red-hot iron with his bare feet. So did Edward Barnwell, who first appeared as the "Human Volcano" in San Francisco in 1879. These better-known performers were succeeded by scores of entertainers willing to suffer ordeals by fire for the sake of vaudeville acts, medicine shows, and circus performances across the country.

They were often called "salamanders," in honor of that creature's mythical ability to endure fire without harm. There was no magic involved, of course, just a few well-known tricks, many of which had been around for centuries. These included coating the body with special ointments and using specially designed safety equipment. Early in the nineteenth century respected scientists like David Brewster went to great lengths to expose the deceptions of the performers; but toward the end of the century a democratic spirit had taken over, and the public was offered directions rather than denouncements. Edward Barnwell described more than fifty fire tricks in his book, *The Red Demons, or Mysteries of Fire;*[47] others in the business offered their own insights. By the turn of the century mail-order companies made it possible for anyone to become a fire-eater. Twenty-five cents sent to Bates and Company of Boston would procure a harmless kit that would enable customers to breathe fire and send brilliant sparks from the mouth "to the horror and consternation of all beholders."[48] That people responded to such offers is clear from a variety of contemporary sources, none more intriguing than a photographic postcard that was mailed from Salisbury, Missouri, in 1907. Scrawled across the image of the town bandstand is the curious message, "This is the place where you saw the little negroes with fire in their mouths."[49]

Of course, fiery effects that could be created by individuals were just that—one-man shows.[50] Bigger effects required bigger operations, and theatrical companies had both the resources and the incentive to put on lavish displays. Burning buildings, exploding machines, and even executions by burning at the stake were standard climaxes of the wildly popular melodramas of the second half of the century. Fire also appeared on stage simply for enhanced atmosphere and excitement. New York City's Yiddish theater offered a good example of this technique in the 1891 production of Joseph Lateiner's four-act play, *The Persecution in Russia*. Following a tumultuous succession of scenes detailing the evil dealings of the tsar and his henchmen, there came a totally superfluous climax consisting of a parade through New York City. Overblown spectacle reigned supreme as actors wearing red, white, and blue regalia marched to the patriotic music of a brass band. Then, as the *New York World* reported, "a tableau is formed: the Stars and Stripes wave side by side with the scarlet socialistic banners, red fire is burned at the wings and the curtain comes down amid a whirlwind of applause."[51]

"Applause" is the key word here, and it is significant that the "whirlwind" came right after the lighting of the red fire. Although many nineteenth-century patrons avoided the theater because of the danger of fire, countless others flocked there to enjoy the spectacle of smoke and flame on the stage.[52] The effects were readily rendered. To suggest mine explosions or military battles, for example, performers used a "quick match," which consisted of powdered sugar and potash mixed together and lighted by a cotton-tipped rod dipped in sulfuric acid. Another useful prop was the lycopodium torch. To make such a device, yellow lycopodium powder was placed in a cup that was connected to an alcohol torch by a yard-long piece of tubing. By blowing the powder into the alcohol flame, a stagehand could produce a brilliant flash that was perfect for imitating burning buildings and fire-eaters.[53] Red fire, a pyrotechnic concoction that produced a glowing red light by burning lithium or strontium salts, was used to highlight almost anything.

The use of special fire effects on stage, as promoted in this advertising card, was a sure crowd pleaser.

Such formulas sound fairly complicated as well as dangerous—best left to the experts, we might say today. One hundred years ago, however, almost everyone was an expert, and fiery displays were produced as easily off the stage as on by an enthusiastic public that often seemed to define a good time solely by the amount of fire power involved. Many indoor occasions were enhanced by the decorative and ceremonial use of fire, but it was outdoors at night that the "controlled conflagration" came into its own.

Open-air gatherings after nightfall afforded the perfect opportunity to exploit the dramatic effects of fire, as Frances Trollope discovered when she attended a camp meeting in Indiana in 1829. Reaching the site about midnight, she first encountered a circle of tents, through which numerous firelights flickered. Then she came upon a scene unlike any she had ever witnessed: "Four high frames, constructed in the form of altars, were placed at the four corners of the enclosure; on these were supported layers of earth and sod, on which burned immense fires of blazing pine-wood." The impact on Trollope was extraordinary. The "lurid glare thrown by the altar-fires on the woods beyond did altogether produce a fine and solemn effect, that I shall not easily forget."[54]

In search of "effect," solemn or otherwise, bandsmen also made forays into the night. Sometimes they marched by torchlight or gathered at a brightly illuminated bandstand, but by far their most popular nocturnal activity during the last quarter of the nineteenth century was serenading their neighbors. Because it was cumbersome and slightly unnerving to tote the big brass instruments through the darkened countryside, bands often recruited torchbearers to light their way. For a much more spectacular effect, however, band experts provided the following directions:

> Provide two or three small fire baskets of hoop-iron, or strong wire, and mounted upon the ends of long sticks. . . . Then during the day preceding the serenade, prepare a number of "fire-balls," made by wrapping a ball of compressed cotton about as large as ones fist with binding wire, and saturating the whole with coal oil by immersion in that liquid for several hours. These balls may be carried by an atten-dant in a tin bucket.
>
> After stealing quietly up to the house to be serenaded and having the men arranged, the music distributed and all ready to begin, place a "fire-ball" in each basket and apply a match. In an instant the whole place will be in a blaze of light, while simultaneously with the burst of flame the signal is given and the music begins.

These procedures usually resulted in a fine show, but bandsmen were warned of potential hazards. A serenading party should "guard against smoking the walls or pillars of a gentleman's house or setting fire to it with their torches," advised one expert in a manual published in 1875.[55]

By this time, bandsmen as well as most members of the public should have been thoroughly familiar with the dangers of torches, fireballs, and other incendiary devices. They had been using them for generations in what was surely the most fire-filled of all secular occasions: the political celebration. The connection between fire and politics had been made long before the settlement of the American colonies. Some European tradi-tions, such as the lighting of bonfires on Guy Fawkes Day, were practiced in the colonies during the eighteenth century. With the coming of the Revolution, Americans scooped up the fireball and ran away with it.

To celebrate the ratification of the Declaration of Independence, exu-berant patriots set bonfires and placed candles and other illuminations in their windows. Thereafter, such displays were repeated annually in many communities to mark the Fourth of July. The bonfire, in fact, became so widespread and so intimately tied with Independence Day revels that it probably should be viewed as the American version of Europe's traditional Midsummer fires.

Other combustibles were added to the arsenal. These included fireballs and a wide assortment of prepared pyrotechnics, among them Roman can-dles, red fire, rockets, and crackers. The illuminated transparency, all but forgotten today, was introduced to America in the eighteenth century and remained in use until the end of the nineteenth century. In Charles

Fireworks and illuminated scenes enlivened a political rally held in New York City on
October 13, 1860. Note the circle of torches around the horseman in the center.
From *New York Illustrated News*.

Willson Peale's version, designed to commemorate the end of the Revolution, a giant transparent painting of the goddess Peace was to be set atop
a building, where she was to appear suddenly from the darkness as lamps set
in the clouds at her feet were lit. Then the transparency was to slide down
a rope to a dark arch below, which, at the goddess's touch, would burst into
light as hundreds of rockets exploded in the air. Far simpler versions of this
device were used to depict candidates' images, particularly in the years just
before and after the Civil War.

Any one of these devices could be used effectively to celebrate national
holidays, campaign rallies, election victories, and other political gatherings. For political torchlight processions, which became immensely popular after the Civil War, the whole arsenal was desirable. One old-timer
recalled:

> In these parades numberless lighted transparences, revealing the
> sentiments of opposing political factions, were borne aloft, and as
> the band, in passing, played inspiritingly and the local drum corps

Fire was a prominent attraction during a nighttime parade held at
St. Louis's Autumnal Festivities, 1886.

drummed thrillingly, they were greeted with unrepressed cheering
from those gathered on the doorsteps and balconies of houses, the
windows of which were illuminated with scores of flickering candles,
while red fire burned at irregular intervals on the curbstones of the
sidewalks along the line of march and brilliantly tinted sputtering
stars shot fitfully into the air from numberless roman candles.[56]

Large municipalities had the resources to stage magnificent displays.
New York City's centennial celebration of the Fourth of July in 1876 could
hardly be surpassed, with its brilliantly illuminated buildings and parks
(Tiffany and Company alone provided fifteen hundred lanterns), its pyrotechnic emblems, and its monumental torchlight procession, which began
at midnight and included more than twenty-five thousand marchers. The
important point about these fire festivals is not their size, however, but
their universality. They were in essence folk festivals, and virtually all
citizens had the freedom, knowledge, and equipment to participate if they
so desired—which, of course, they did, as the following newspaper report
suggests. The year is 1888, the place is a small mill town in rural Rhode
Island, and the occasion is the Republican victory in a recent election:

On Tuesday evening the Republicans of this village and Wyoming
held a jollification in honor of the Republican victory. A tower of tar
barrels was constructed on the hill near the Republican flag, reaching
high in the air, and while burning made a very attractive feature. The

Republican members of the old Jillson Cornet Band, now disorganized, came together with some twelve or fourteen pieces and gave a very creditable concert on the hill near the bonfire of tar barrels. Captain H. F. True had the heavy cannon from the Machine Company's works in position and kept up a continual cannonade for hours. At the report of the fifth gun a lot of fireworks and red fire were sent off all along the line from this place to Wyoming. About eighty fires were sending up their lurid glare. A drum corps also paraded the streets. The citizens were out in large numbers to witness the illuminations, some of which were very pretty. . . . Mr. Patterson . . . had a bonfire burning when the band and its followers passed his residence on the way to Wyoming, where about the same order of exercises were carried out as in this village, tar barrels ablaze, fireworks, red fire and music.[57]

This "order of exercises," standard in most villages at the time, was good entertainment, and yet it was also something more. In a curious transformation of the traditional Easter fire festival in which individual householders took a brand from a central, consecrated fire to light their own family hearths, the patriotic festivals of the United States called for individual householders to *give* fire to the communal celebration. This was, perhaps, the most appropriate way for a democratic fire festival to operate.[58]

CHILD'S PLAY

In the world of the open flame, fire and play formed an early and long-lasting alliance. A survey of children's toys alone suggests the overwhelming importance of fire in everyday life.

Fire engines probably rank among the most popular playthings of all time. Even before the Civil War, handmade toy steamers lured young children to the nursery just as the real ones lured their elders to the streets. With the establishment of successful toy manufactories in America in the 1870s, the production of such vehicles flourished. Brightly painted hand tubs, steam pumpers, and hose reel carriages were widely available, often accompanied by firemen, horses, and a firehouse. Some were touted as "working" models: bells clanged, wheels turned, and, in the most expensive sets, live steam issued from the boilers.

Other popular toys with a firefighting theme also appeared by the end of the century. Now prized as collectors' items, these objects range from sectional puzzles depicting firemen racing to a fire to fancy dress-up kits consisting of regulation-style hat, axe, red belt, breastplate, and speaking trumpet.[59]

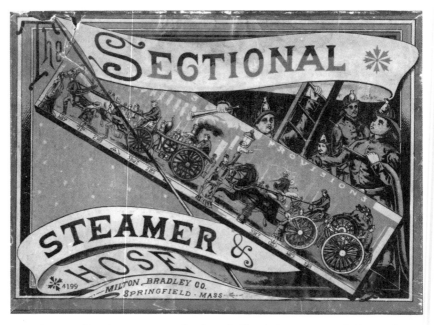

Milton Bradley Company capitalized on children's fascination with firemen by selling toys with firefighting themes. This thirty-six-piece sectional wooden puzzle features a variety of fire engines.

Whether constructed of tin, wood, cast iron, or pressed cardboard, most of these toys were carefully crafted to appear realistic. That was also the idea in the miniature world of the "baby" or doll house, where, in the best examples, the fire technology used in real houses governed the design of everything from chimneys to mantels and hearths. One exceptional early American doll house, probably built in Philadelphia around 1810, even has a pair of Green Tree fire marks on it.[60] But it is in the furnishings that manufacturers excelled, offering such utilitarian items as andirons, fire tools, fire grates, Franklin stoves, parlor stoves, hinged coal boxes (with coal), and lead filigree chandeliers, which, if given a little fuel, would actually light. Kitchen equipment was similarly lifelike: cast-iron cook stoves came complete with dampers, lifts, pipes, and cooking pots. By the end of the century such play stoves had become immensely popular toys even without the accompanying doll house. They were widely available in "perfect working" models from such mail-order houses as Sears, Roebuck and Company for as little as fifty cents.[61]

The Sears advertisements used descriptions like "perfect working" for a reason. As the company well knew, children had a strong interest in handling fire; and, like the adults who lived and worked around them, they would often go to great lengths to incorporate fire into their recreational activities.

Growing up in antebellum Boston, Edward Everett Hale and his friends often enjoyed making expeditions to nearby woodlands. "Phosphorous-boxes," those magical flame-producing kits just coming into vogue, were at the top of the list of necessary equipment. "When we arrived at the woodland sought," Hale explained, "we invariably made a little fire. We never cooked anything that I remember, but this love of fire is one of the earlier barbarisms of the human race which dies out latest. I suppose if it had been the middle of the hottest day in August we should have made a fire."[62]

Half a century later and almost three thousand miles to the west, Sarah Bixby came to the same conclusion. She describes in her diary a memorable occasion when "the spirit moved me to cart a shovelful of live coals out through the door to the porch, and there coax up a fire by the addition of kindling wood." Fortunately, another more benign spirit urged the little fire starter to call her mother to see her "nice fire." Disaster was averted, but not Bixby's interest in flames. "Fire, candles, matches, revolvers all held a fascination," she confessed.[63]

As if to accommodate this attraction, adults created—or at least sanctioned—a startling number of fire-related pastimes for children. Today, when fire play is limited primarily to birthday candles and the occasional Independence Day sparkler, it is almost impossible to comprehend the permissiveness of the last century.[64] Children played with sealing wax, Roman candles, and toy steam engines that burned alcohol as fuel. They experimented with magnifying or "burning" glasses, readily using the sun's rays to set fire to dry leaves, shavings, and—in Hale's case—gunpowder. Among the "instructive and profitable" toys enjoyed by affluent youngsters at the turn of the century were the magic lanterns sold by Sears and other firms. These entertainment outfits, which included the lantern, slides, advertising posters, and tickets, supposedly taught children how to run a business. What advertisers failed to state explicitly was that the kits also taught children how to fill and light the kerosene or coal oil lamps that generated the flame to project the images.

Homemade entertainments could be just as amusing and just as flamboyant as the store-bought variety. Daniel Beard's *American Boys Handy Book* (first published in 1882) provides an intriguing window on the world of fire play at the end of the nineteenth century. As might be expected from a man involved in the establishment of the Boy Scouts of America,

"Sea-Side Sketches—A Clam-Bake," after Winslow Homer. These boys are clearly at ease
with the manipulation of fire. Note the boy to the left carrying fuel.
Published in *Harper's Weekly*, August 23, 1873.

Beard's compendium of recommended pastimes includes long sections on
camping skills. With practical advice and informative diagrams, he taught
his young readers exactly how to turn a burnt barrel into an oven and how
to build a fire, rake the coals, and cook a meal—all without the assistance
of an adult. For indoor play, Beard suggested making smoke rings and turning a human finger into a match. The latter activity was simple enough:
"To light the gas with your finger, turn it on without applying a light, walk
around the room, sliding your feet over the carpet, until you again reach
the burner, touch the tip of the burner and instantly the light will blaze up
as if by magic."[65]

Not content to let the boys have all the fun, Beard's sisters, Lina and
Adelia, decided in 1887 to create their own handy book for girls. Even
though girls were more likely than boys to need fire-making and cooking
skills, camping procedures were not included in their version. They did not
neglect the open flame in their roster of amusements, however. The following party games were recommended for Halloween, an occasion traditionally celebrated with fire:

THE APPLE AND CANDLE GAME

From the ceiling is suspended a stout cord, the lower end of which is securely tied to the centre of a stick about a foot and a half long. On one end of the stick is fastened an apple, on the other a lighted candle. The string is set in motion, swinging back and forth like a pendulum, and the contestants for the prize stand ready, each in turn, to make a grab for the apple, which must be caught in the teeth before it can be won. Frequently the candle is caught instead of the apple, which mishap sends the spectators off into shouts of merriment.

A LIGHTED CANDLE

The candle is placed upon a table in full view of everyone; then one of the players is blindfolded, turned around several times, and set free to seek for the candle and blow out the light, if possible.

THE GHOSTLY FIRE
[recommended only for girls with "strong nerves"]

Salt and alcohol [are] put in a dish, with a few raisins, and set on fire. As soon as the flame leap[s] up we clasp hands and gayly danc[e] around the table, upon which burn[s] our mystic fire. . . . The dance [is] not prolonged, for it [is] our duty, before the fire [is] spent to snatch from the flames the raisins we had put in the dish. This can be done, if one is careful, without as much as scorching the fingers.[66]

For the Fourth of July, both handy books came up with fiery entertainments that make the Halloween diversions look like parlor games. Girls were shown how to make illuminated paper tents filled with candles. They could also construct an elaborate holiday decoration by tying lighted Chinese lanterns to the spokes of an umbrella. For the boys, there were fireworks, many designed to explode in a shower of colorful sparks. Simple instructions told the reader all he needed to know to transform a Roman candle containing colored balls, a piece of wire, and fuse into a beautiful pyrotechnic:

The fuse used can be bought in almost any city or town; it is sold to miners for setting off blasts. With the wire make a sort of wheel, with two or three spokes; cut open the Roman candle and extract the powder and balls; wrap up each ball with some of the powder loosely in a piece of tissue-paper and tie the paper at the ends upon the spokes or cross wires of the wheel. . . . Run the fuse spirally around, passing through each parcel containing a ball, and allow the long end of the fuse to trail down beneath from the centre or side.

Then all a boy had to do was connect the wire wheel to the mouth of his balloon and light the fuse as the balloon rose in the air. In short order, the fire should reach the balls, igniting them one by one and causing "showers of pretty jagged sparks." "By the time all is over," promised Beard, "the tiny light of the solitary ball in the balloon looks like a star in the sky above."[67]

In suggesting these activities, the Beards betray not so much a tolerance of fire play as a hearty endorsement of it. In such a climate, it is hardly surprising that children picked up where their elders left off and invented their own methods of playing with fire. For one midwestern boy this led to lighting a few shavings at the fireplace and setting his brother's bushy head of hair on fire.[68] For a California girl, the motivation was more sentimental. Seeking to recapture the joy of Christmas from the preceding day, she stole some tapers from the Christmas tree and lighted them in a make-believe tent made from bedclothes. As an encore, she took her new toy stove, stuffed it with paper, and set it on fire, presumably to enjoy another holiday feast.[69]

William Dean Howells and his friends channeled their energies into the political arena and favored election night fireballs and bonfires. For the former, they took wads of cotton, soaked them in turpentine, and set them on fire. Then, while the balls were blazing wildly, they scooped them up and threw them high into the night air. Bonfires required less nerve but more muscle. After making a stack of tar barrels as high as the tallest boy could reach, they lit the tower with a match. Moments later, the celebrants would be rewarded with a dusky, smoky flame that waved from the top "like a crimson flag." This was the signal to dance around the tar barrels and cheer the Whig party. Howells admits that the boys had to be nimble to avoid getting burned. Nevertheless, the town survived the annual conflagrations, and the boys were left undisturbed in their celebrations.[70]

To Howells, such freedom only confirmed what he already knew, namely, that Hamilton, Ohio, was "a glorious town . . . for boys." Other places were not nearly so lenient. Fun with fire could lead to serious injuries: scientist Benjamin Thompson was sent home from his apprenticeship after being injured making a rocket; Henry James's father lost a leg in a fire-balloon accident; and Daniel Beard himself was almost blinded by red fire.[71] Many jurisdictions responded by enacting strict laws. Newport, Rhode Island, signaled the trend by prohibiting fireworks within the city limits early in the eighteenth century.[72] Other places focused on the problem of children and fire, as when, in 1854, residents of Greenfield, Indiana, forbade the "shooting [of] fire-crackers or sky-rockets, by any boy or children under the age of twenty-one years, except the parent or guardian be present with him or them at the commencement and during the whole time of said burning."[73] Less binding, but equally sincere, were the

WORK AND WIN

An Interesting Weekly for Young America.

Issued Weekly—By Subscription $2.50 per year. Entered as Second Class Matter at the New York Post Office, December 8, 1898, by Frank Tousey.

No. 336. NEW YORK, MAY 12, 1905. Price 5 Cents.

FRED FEARNOT AS A FIREMAN;
OR, THE BOY HERO OF THE FLAMES.
BY HAL STANDISH

"Here! Here!" cried Fred. "You must get out or you will be burned up." "Save him! Save him!" cried the young girl. "He is chained to the wall!" Fred saw the chain, and sprang forward and cut it loose with his axe.

Fred Fearnot stories were a regular feature of *Work and Win*, a boys' magazine of the turn of the century. The exploits of firemen appealed enormously to the youthful readership.

countless warnings of danger that appeared in both juvenile and adult publications throughout the nineteenth century. "As we commemorate [on July 4] the escape from the dreadful effects of gunpowder," intoned one, "it is rather absurd that it should on that day be made the principal agent for amusement."[74] The movement for "daylight" or "harmless" (that is, noncombustible) fireworks, which both the girls' and boys' handy books applauded, was another response to the problem.

And yet, there was something incomplete, almost half-hearted about these efforts prior to the twentieth century. Many adults may simply have considered fire play among children to be inevitable. And perhaps that attitude was perfectly proper in a society where fires burned at every turn. Play was, at the very least, a practical way to teach youngsters about the complex nature of fire in their world. The well-known saying "Once burned, twice shy" was preceded by a sixteenth-century version: "A burnt child dreads the fire." Both dictums imply that experience is the best teacher.

William Dean Howells was one youth who learned the lesson well. He discovered at an early age that, when respected, fire offered not only energy but also recreation and inspiration. More exciting still, fire offered these things to boys. These ideas find expression in Howells's evocative descrip-tion of the caves of his childhood. As he remembered it, the boys built these magical caves

> with a hole coming through the turf, to let out the smoke of the fires they built inside. They had the joy of choking and blackening over these flues. . . . They never got so far as to parch the corn or to bake the potatoes in their caves, but there was the fire and the draft was magnificent. The light of the red flames painted the little, happy, foolish faces . . . as the boys huddled before them under the bank, and fed them with the drift, or stood patient of the heat and cold in the afternoon light of some vast Saturday waning to nightfall.[75]

The words are Howells's, but the love of fire would have been recognizable to most all his contemporaries.

Advertising Magic

Fire was an advertiser's dream, particularly in the second half of the nine-teenth century, as advertising blossomed with the country's gradual trans-formation into a consumer society. As might be expected, fire was thor-oughly exploited to sell anything relating to fire safety and protection. Insurance companies placed images of burning buildings on their adver-tisements. Manufacturers of fireproof safes published dramatic testimonials

The domestic hearth was used to advertise Parker's Hair Balsam in
this trade card, ca. 1870.

from customers whose safes had survived hours of red-hot fire. Purveyors of
fire extinguishers even built fires to demonstrate the effectiveness of their
wares, although P. T. Barnum's experience with a malfunctioning Phillips
Annihilator proves that this dramatic approach could backfire. At their
best, however, such tactics succeeded, and the reason is clear: after in-
flaming the public's instinctive fear of raging fire, the advertiser was able
to offer salvation through the purchase of a particular product or service.

Much less obvious is the connection between fire and patent medicines.
Yet fire helped to sell hair balsam, cough syrup, and all-purpose concoctions like the Hamlin Company's Wizard Oil. One favorite ploy was to
include fire-eaters among the acts in traveling medicine shows. Like the
tumblers, magicians, and singers who appeared with them, fire-eaters attracted people's attention so the salesman could make his pitch. Medicine
manufacturers also used fire to reach audiences via print. In its magazine,
the Kickapoo Indian Medicine Company included among its advertisements such bizarre stories as the dreadful tale of outlaw Johnson Harris,
who "perished at the stake, roasted alive!" The sensational illustration of
the stake scene that accompanied this dramatic tale undoubtedly received
more study than the adjacent testimonial by Gertrude Wallbank.[76]

Companies with smaller ambitions and budgets often chose to distribute
lithographed trade cards instead of magazines. These, too, often depicted

Old Homestead Brand Apples employed fireside symbolism
in promoting their product.

fire, either because the bright red color was conspicuous or because a useful
connection could be established between, for example, a hearth fire and a
"family" remedy. Between 1880 and 1900, countless commodities in addition to medicine were touted on colorful trade cards with an associated fire
motif.

In fact, the possibilities for pairing products with fire imagery were seemingly endless. Hearth fires sold Old Homestead apples, and factory fires
sold McKinley's presidential candidacy. Illustrations of conflagrations sold
newsweeklies, and descriptions of fires helped sell books. Sometimes fires
even sold towns. When railroad companies wanted to attract settlers to
new communities along their lines, they often published bird's-eye views
of the streets and buildings—complete with smoking trains, steamboats,
and chimneys. Occasionally artists even enhanced reality by including
more industrial smoke than local factories actually produced, just to under-
score the community's prosperous future.[77]

Selling an idea was more difficult than selling a product, but fire met this
challenge too in flaming colors. Probably the most obvious use of fire in
this context involves what we might call the "scared straight" approach.
By invoking a fear of fire—particularly fear of bodily injury—a zealous
reformer could manipulate the behavior of others. The most obvious example of this hard-sell technique is the Puritan vision of hell. The words
"fire and brimstone" roll off our tongues today as if they were nothing more
than the brand name of some unpleasant medicine, but to generations

St. Louis in 1846, an oil painting by Henry Lewis (1819–1904), uses the billowing
smoke of steamships and chimneys to evoke a scene of bustling progress.
Courtesy of the St. Louis Art Museum.

of Americans this phrase conjured up visions of eternal torture amid
scorching flames.[78] The notion that hell was both real and horrifying was
axiomatic by the seventeenth century, and clergymen like Thomas Shep-
ard had no difficulty finding the words to describe it. "God shall set himself like a consuming infinite fire against thee," he warned all backsliders.
"Thou canst not endure the torments of a little kitchen-fire, on the tip
of thy finger, not one half hour together. How wilt thou bear the fury of
this infinite, endless, consuming fire, in body and soul, throughout all
eternity?"[79]

The same technique was still in use decades later. In his aptly titled
tract, *The Efficacy of the Fear of Hell to Restrain Men From Sin* (1713),
Solomon Stoddard solemnly warned his flock that if they did not adopt a
godly life they would be damned to hell, where "the Worm dyeth not . . .
[and] . . . the Fire is not quenched."[80] "O sinner!" Jonathan Edwards exclaimed to a gathering of the faithful in Enfield, Connecticut, in 1741;
"Consider the fearful danger you are in: it is a great furnace of wrath, a wide
and bottomless pit, full of the fire of wrath, that you are held over in the

The view of the earth's interior as a scorching, fiery region was amplified by many preachers, who equated volcanic destruction to God's wrath against sinners. Peters Collection, National Museum of American History, Smithsonian Institution Photo 60416-B.

hand of that God, whose wrath is provoked and incensed as much against you, as against many of the damned in hell. You hang by a slender thread, with the flames of divine wrath flashing about it, and ready every moment to singe it, and burn it asunder; and you have no interest in any Mediator, and nothing to lay hold of to save yourself, nothing to keep off the flames of wrath."[81]

Such terrorism was beginning to seem excessive to many clergymen even before Edwards's day. Yet, even as Protestant theology de-emphasized divine punishments, fear of hellfires persisted, amplified from time to time by religious revivals and the ecstatic exhortations of fundamentalist preachers. If surviving autobiographies are any indication, countless American children growing up in the decades before the Civil War suffered from thoughts of being incinerated.

By this time, however, fire was being used to champion a new idea: social reform. One example, taken from the temperance crusaders of the 1820s, presents a strangely familiar evocation of fire. In the following excerpt from his *Address on the Effects of Ardent Spirits*, Jonathan Kittredge voices a commonly held belief about the combustibility of heavy drinkers:

> I read of an intemperate man, a few years since, whose breath caught fire by coming in contact with a lighted candle, and he was consumed. At the time I disbelieved this story, but my reading has furnished me with well authenticated cases of a combustion of the human body from the use of ardent spirits. . . . I will state one of them, and from this an idea can be formed of the rest. It is the case of a woman eighty years of age, exceedingly meagre, who had drunk nothing but ardent spirits for several years. She was sitting in her elbow chair, while her waiting maid went out of the room for a few moments. On her return, seeing her mistress on fire, she immediately gave an alarm, and some people came to her assistance, one of them endeavored to extinguish the flames with his hands, but they adhered to them as if they had been dipped in brandy or oil on fire. Water was brought and thrown on the body in abundance, *yet the flame appeared more violent, and was not extinguished until the whole body was consumed.*[82]

The destructive side of fire was clearly a powerful tool of persuasion. It is not inconceivable that in skilled hands fire could have been used against itself, scaring people to abandon combustion as their energy source. And yet, it was the gentle side of the flame that helped to sell one of the most fundamental and influential notions of the nineteenth century: the sanctity of the home.

That there was a connection between fire and domestic well-being was an inescapable conclusion in the days of open-hearth cooking and heating. During the eighteenth century, Benjamin Franklin expressed the idea matter-of-factly with a pert almanac aphorism: "A house without woman and firelight is like a body without soul or sprite." But a curious thing happened to this concept on the way to the twentieth century. Caught up in the whirlwind of domestic and technological changes that buffeted American society during the 1800s, the family hearth gradually shed some of its more mundane attributes in order to take on a potent new symbolism. It became, in essence, the altar of a curious new religion that transformed the American house into a temple of secular virtue and elevated the wife and mother to the role of high priestess and defender of the flame.

Compliments of
RAUCH, THE HATTER,
No. 713 Girard Avenue.

A fire in the hearth symbolized
a happy home. Trade card,
ca. 1870.

The celebration of the family home as a tranquil haven from the crass competitiveness of the outer world began early in the nineteenth century. Major contributors to this "cult of domesticity" were women, clergymen, and legions of hard-working husbands who valued the idea of a refuge from toil. There was no particular reason that fire had to figure in this scheme, but most writers on the subject viewed the hearth as the heart of the home and therefore used the fireside as a rallying point. In 1864 the English writer John Ruskin captured the essence of this philosophy:

> This is the true nature of home—it is the place of peace; the shelter, not only from all injury, but from all terror, doubt and division. In so far as it is not this, it is not home; so far as the anxieties of the outer life penetrate into it, and the inconsistently-minded, unloved, or hostile society of the outer world is allowed by either husband or wife to cross the threshold it ceases to be a home; it is then only a part of the outer world which you have roofed over and lighted fire in. But so far as it is a sacred place, a vestal temple, a temple of the hearth . . . it is a home.[83]

This idea was incorporated into American art and literature in countless ways. Artists depicted mothers teaching their children in the ruddy glow of the fireside, poets described "Home Sweet Home," and Christian educators promoted the hearth as a holy shrine. Writers of fiction also glorified "the blazing wood fire" as the "household altar," and some went on to publish their work in such appropriately titled magazines as *The Happy Home and Parlor Magazine* and *Hearth and Home*.[84]

As the reminiscences of ordinary people make clear, the concept took hold in private lives as well. A New Englander named Levi Hutchins was about to depart for home one evening in 1810 when he was dissuaded from travel by the irresistible allure of his brother's hearth: "The social circle of my brother's household, cheered by the mingled light of the bright woodfire and his domestic tallow-candles, caused so much happiness, that I was induced to postpone our return till morning."[85] Much later in the century, the equation of the hearth with domestic stability and comfort had become a virtual synecdoche of secular American thought. Jonathan Dawson, whose family had settled in Indiana in 1837, produced a typically reverential description around 1910. Asked to recall the rugged homestead for a newspaper article, he dutifully described the log walls, the plank roof, and the oiled-paper windows. "But," he added, "it is of the wide cheery fire place, around which we delighted to gather, and in whose ruddy glow cluster the most sacred memories of my childhood, that affords me the greatest pleasure to describe."[86]

If ever there was a successful advertising campaign, this was it.

Firing the Imagination

Come ye, who form my dear fireside—
My care, my comfort and my pride;
Come now, let us close the night,
In harmless talk, and fond delight.[87]

It is such "harmless talk" that constitutes one of the open hearth's most cherished—and least tangible—legacies. Fire could entertain, but, more than that, it could stimulate the creation of new entertainments as well. The possibilities were limited only by the imaginations of the people gathered at the fireside.

Fiction of the nineteenth century abounds with references to the phenomenon of fireside gossip and storytelling. Harriet Beecher Stowe's *Oldtown Folks* devotes a full chapter to the pleasure of "fire-light talks in my grandmother's kitchen." Nathaniel Hawthorne also alludes to the chimney corner tradition, though his treatment is more somber and allegorical, stressing the mysterious filtering of ancient family lore through the "smoke of the domestic hearth."[88] Taking off on a slightly different tangent, some authors of the day became so enamored of the concept that they let the andirons or the fire screen themselves tell intriguing domestic tales.[89]

The interior of Samuel Mount's cabin was portrayed by John Collins in the 1870s. The blazing hearth is clearly an impetus to conversation and conviviality. Courtesy of the Friends Historical Collection, Guilford College, Greensboro, North Carolina.

As contemporary diaries and memoirs make plain, this literary tradition grew out of a widely practiced folk genre. "We are seated as thick as bees around the kitchen fire everyone putting in a word to keep up the story," wrote a Connecticut schoolgirl named Delia Hoyt to her parents in 1825. "Grandfather and old Aunt Nabby are chatting their youthful tales— Grandmother is talking about you. Robert is reading Mary and Nelson are eating apples Rufus and Abby in the corner and Mary Williams ironing— so you can imagine how we look round the fire with a plate full of excellent apples in the middle."[90] Later in the century, Lucy Larcom described similar occasions from her youth: "When supper was finished, and the tea-kettle was pushed back on the crane, and the backlog had been reduced to a heap of fiery embers, then was the time for listening to sailor yarns and ghost and witch legends."[91]

If other "ear-witness" accounts are accurate, Americans also engaged in enthusiastic debates on religion and politics or shared spicy tidbits of neighborhood news interspersed with the superstitious sayings of days gone by. Washington Allston, the romantic painter of the early nineteenth century, "delighted in being terrified by the tales of witches and hags" that slaves told him by the fireside of a plantation cabin.[92] Virginia Wilcox

Ivins, an early immigrant to California, preferred romantic songs around a campfire. Andy Adams, a cowboy turned author, refused to limit his yarning and boasted that his own campfire tales ran the gamut "from the sublime to the ridiculous, from a true incident to a base fabrication, or from a touching bit of pathos to the most vulgar vulgarity."[93] Clearly, fireside stories were as diverse as the firesides that generated them.

Publishers were quick to capitalize on the phenomenon. Any number of volumes were propelled into print with the magical word *fireside* as part of the title. The tag was an automatic encomium, and as open hearths began to disappear it could be found increasingly on treatises ranging from religious education and biography to children's collections of stories and songs. In content, too, many books drew heavily on the fireside tradition. In 1858 the Philadelphia firm of Lindsay and Blakiston brought out a history of the American Revolution in which the action was recounted by Continental soldiers sitting around campfires in the various theaters of war. But, like the presidential fireside chats of the twentieth century, these many publications simulated the pattern of a fireside gathering without capturing its underlying spirit. That spirit required the presence of the open flame.

To say that the flickering of a flame, the dancing of shadows on the walls, and the mysterious movement of warm and cool air across one's body are elements that foster creativity is, perhaps, to say too much. The concept has had its proponents, however. Samuel G. Goodrich, a prolific writer of nonfiction for children, prefaced his 1849 survey of the industrial and artistic achievements of mankind by acknowledging the contribution of his parlor fire to the book's genesis and format: "I was dozing by my evening fire-side, when one of those hasty visions passed before my mind, which sometimes seem to reveal the contents of volumes in the space of a few seconds. It appeared as if every article of furniture in the room became suddenly animated with life, and endowed with the gift of speech; and that each one came forward to solicit my attention and beseech me to write its life and adventures."[94] August Kekule claims to have had an analogous experience, except that in the case of this thoughtful scientist the motion of the flames conjured up images of atomic arrangements and, ultimately, the configuration of the benzene ring. Ever the pragmatist, Benjamin Franklin gazed into *his* fire and saw a way to improve fuel efficiency.

One has to look no farther than the fireside story to see that inspiration was fire's gift to people of less purposeful outlook as well. Somehow, a group of convivial people, a darkened room, and a glowing hearth were the only tools needed to fire the imagination and produce home-style entertainment in households across the country. For Ellen Rollins, the "fiery tongues" of the backlog of her childhood fireplace were so eloquent that "the stories it told to a child, with crackling voice, went not out with its smoke."[95] Henry Wadsworth Longfellow expresses similar sentiments in "The Fire of Drift-Wood," a poem in which the "bickering flames" of a driftwood fire conjure up memories and talk of times gone by.[96]

Stories could be told without the benefits of the flame, of course. Nevertheless, it seems to have been the almost universal belief during the nineteenth century that the energy emanating from an open fire helped the human spirit—as well as the human body—to do its work. Conversely, many also believed that the loss of the flame meant the loss of a cultural phenomenon. Lucy Larcom makes the point emphatically: "The wonder seems somehow to have faded out of those tales of eld since the gleam of red-hot coals died away from the hearthstone. The shutting up of the great fireplaces and the introduction of stoves marks an era."[97]

Larcom may have been somewhat premature in her eulogy; but it is clear that by the time central heating had become the norm, the fireside tradition was well on its way to becoming a periodic pastime for campers. Nor was it the fireside tradition alone that disappeared; in the absence of recording devices, the words that it created are largely gone too. We can get some sense of the range of topics for fireside tales, but the use of language, the pace of the stories, the give-and-take of the participants are lost.

Historic restorations can tell us many things about fire—the size and shape of the hearth, the tools that were used there, the range of work that was done. But even with the most rigorous research, there are some uses of fire at which we can only guess.

Bad Master

On the evening of September 19, 1902, two thousand people elbowed their way into the Shiloh Baptist Church in Birmingham, Alabama, to hear a speech by the renowned educator Booker T. Washington. But what began as a joyful celebration of Baptist theology and African-American culture suddenly turned to tragedy when, at about nine o'clock, the cry of "Fire!" cut through the crowded hall. Scores of men and women began an unruly and increasingly hysterical retreat to the doors. Despite repeated requests by the ministers for order, the stampede gained momentum. By the time fire engines arrived, the scene was a nightmarish vision of bodies piled ten feet high near the exits. The firefighters went to work immediately, but before the situation could be brought under control, more than one hundred people had died trying to escape. The only aspect of the affair more horrifying than the casualty count was the astonishing fact that there was never any fire at all.

Clearly, the members of the congregation had learned the sober truth about combustion. Fire, the fountainhead of domestic comfort and the energizer of American industry, could also be a devastating destroyer. Its reputation as a ravager of crowded assemblies was so entrenched that as soon as the Shiloh audience mistook mumbled reports of "fighting" for the cry of "fire!" there was no stopping the panic. Similar incidents had occurred in the past,[1] inspiring not only fear among theatergoers but also Oliver Wendell Holmes's now-famous declaration that free speech does not include the right to yell "Fire!" in a crowded theater.

Both responses were understandable. Fire raging through an auditorium was known to be a merciless antagonist. As early as 1811, a lighted candle swaying carelessly against some lightweight scenery panels ignited the highly flammable interior of the Richmond Theatre in Virginia. Within minutes, the entire building was ablaze and at least seventy people had perished. Less lethal, but no less swift in its destruction, was the 1838 fire that devoured Baltimore's Low Street theater along with its splendid new set of scenery. New York City, home to numerous theatrical troupes, suffered chronically from the problem. The Brooklyn Theatre fire, which claimed almost three hundred lives in 1876, was only the most notorious

The drama depicted on this unidentified stereograph has been replayed thousands of times across the country during the last two hundred years.

of the thirty or so serious theater blazes that tortured the metropolis during the nineteenth century.[2]

For vulnerability to fire, the outside world was every bit the equal of the stage. Out-of-control blazes routinely consumed whatever happened to be in their paths—paths that, as often as not, veered suddenly and unpredictably in response to the weather and the nature of the fuel. Fire has been called the great transformer, but its awesome power has always included the fearful ability to turn communities into rubble, art works into ashes, and wealthy citizens into paupers. The popular press, ever ready to report on human misfortune, habitually called attention to the destructive side of fire by describing it as a "fiend," a "red demon," and "the ravenous element." To one Cincinnati editor writing in 1807, fire was quite simply the "most destructive of all elements."[3] This opinion gained considerable supportive evidence over the next hundred years as numberless fires laid waste to American cities, businesses, factories, and homes. So ruinous has been the legacy of the open flame that it has been observed that if fire were discovered today it would not be approved by the government for general use.[4]

The ancient Romans incorporated notions about this ominous side of fire into their pantheon. Vulcan, particularly in his earlier characteriza-

66

"Burning Charcoal, N.C." This stereograph (half view)
portrays a typical nineteenth-century collier's mound.

tions, represented fire-the-destroyer and thus stood in obvious contrast to
Vesta, the gentle goddess of the hearth. For Americans of the nineteenth
century, a more meaningful symbol of the duality of fire would be a single
individual: the collier.

Today, the word *charcoal* evokes little more than visions of a good bar-
becue. Years ago, however, charcoal was valued as the precious fuel that
fed the blacksmith's fire and helped produce the nation's iron and steel.
Charcoal was manufactured by colliers, a skilled fraternity of workmen
who had an interest in combustion not simply because their product was
destined for oxidation but, more importantly, because the process of manufacture required dexterous manipulation of the open flame. Charcoaling
involved the transformation of ordinary wood into a rich carbon fuel
through carefully controlled burning. Simple heaps of neatly stacked
lengths of wood were covered with a thin shell of leaves and dust, then
ignited and allowed to smolder for several days. Gases escaped out a vent
hole, but additional oxygen was excluded by the protective shell. If all
went well, the batch would be done within a few days, and the collier
could rake the heap apart, cool it, and ultimately sell it. If, however, the
amount of oxygen was insufficient, the product would be incompletely
charred. Worse still, if there was too much oxygen, the entire heap would
go up in smoke and flame. Thus, if a collier failed to control his fire, he
risked enormous loss.[5]

In a sense, America was a nation of colliers—an entire population balanced on the edge of disaster. Throughout the century, Americans willingly, almost eagerly, harnessed fire to their needs even as they constantly ran the risk of being destroyed by it. The thin line between "fire the servant" and "fire the master"—already a tenuous barrier owing to the ease of fire making—was rendered almost imperceptible by the fact that the material world was, in effect, one gigantic fuel heap. This situation had advantages in certain instances. Wooden flatboats, for example, floated commodities down river to market and then were immediately broken up as firewood, thus making a profit for the boatmen and sparing them an arduous return voyage. In extreme cases, wooden implements and furniture were used to stoke the life-saving fires of individuals too sick or too lost to find conventional fuel. But in general, the extraordinary flammability of the average home and workplace during the age of the open flame was acutely problematical.

Danger lurked everywhere. The use of wood as a construction material for everything from houses and barns to sailing ships, bridges, and railroad cars posed the most obvious threat. Then there were the miles of wooden sidewalks erected across the nation to protect pedestrians from the dust and mud of unpaved streets. Similarly vulnerable to accidental ignition were wooden roadways such as the wood-block pavements that became common in the North and Midwest just after the Civil War and the "charcoal roads" that were built sporadically in heavily timbered areas throughout most of the century.[6] Woven materials used in theaters for scenery, curtains, and costumes were consumed as readily as the flimsy fabrics that turned up in most middle-class homes as window dressing, bed furnishings, and attire. William Allen White describes in his memoirs an occasion when a prairie fire was stopped just short of his school's woodpile, the implication being that the disaster would have been much worse had the fire reached the neatly stacked fuel.[7] To a conflagration in full fury, however, mankind's designation of what was or was not fuel meant nothing. A raging fire could burn the dry prairie grasses where they sprouted as easily as in a stove; it could consume living trees almost as quickly as neatly stacked lengths of timber. Ironic instances of fires that burned the right fuel in the wrong place were regularly communicated to the American public via the somewhat sensationalistic reports of coal mine fires that burned unimpeded for years at a time.[8]

Sometimes the combustibility of the basic building blocks of American life remained hidden or ignored until a disastrous fire reminded the public of its perpetual vulnerability. The owners of the Richmond Theatre assumed that the brick exterior would offer some measure of fire protection;

but as the 1811 blaze quickly revealed, the interior canvas walls and the wooden roof had no such fireproof qualities. Much later in the century, residents of Boston were confident that their solid granite edifices would retard the spread of fire. The great fire of 1872 proved otherwise. Stone façades blistered and exploded in the intense heat, and unquenchable flames danced from building to building along highly flammable wooden cornices that were beyond the reach of firemen's hoses.[9]

A further complication arose from the fact that private homes and businesses habitually harbored dangerous substances within their wooden walls. As contemporary accounts make clear, innumerable urban conflagrations intensified the moment they encountered buildings used to store such highly flammable items as whiskey, animal fat, natural gas, and, most devastating of all, gunpowder. It is generally agreed that the fire that struck Savannah, Georgia, on January 10, 1820, would have been controlled fairly easily had it not been for a cache of gunpowder stored in one of the burning buildings. The ensuing explosion instantly escalated the crisis and "struck with terror almost everyone."[10]

The same might have been said of almost every fire that escaped human control. Communities were at risk virtually every time the fire alarm rang because even the smallest spark had the potential for enormous damage. In Hartford, Vermont, for instance, when a passenger train plunged off a bridge during a cold night in February 1887, the sequence of events was appalling. First, the overturned stoves and oil lamps ignited the wooden railroad cars, which were soon ablaze. Fanned by the winter wind, the fire engulfed the wooden trestle bridge, which tumbled in a flaming mass back upon the train and intensified the already raging fire there.[11]

This domino effect was nothing new. In 1791 a fire that began in a Philadelphia livery stable spread so easily along the frame edifices of Dock Street that it consumed fourteen buildings before being brought under control. Three years earlier a candle in a New Orleans dwelling ignited a piece of lace and set off a conflagration that destroyed more than eight hundred buildings. And as is well known, a lamp overturned in a Chicago barn during the dry autumn of 1871 leveled a city.

A mid-nineteenth-century pamphlet depicted fire as a terrible force "carefully watching, night and day, for a moment's inattention in order to sweep the whole into an agonizing . . . grave."[12] Little wonder that Edith White, who spent her childhood in the tinder-dry rustic settlements of mid-nineteenth-century California, had an inbred fear of the fire alarm. She probably spoke for most of her contemporaries when she confessed, years later, that "the sound of that bell in the night fills me yet in memory with palpitation and pallor."[13]

This chapter explores fire's dark, destructive side. Although property loss constitutes the most conspicuous aspect of the subject, generations of Americans suffered in many other ways from their fires.

COMBUSTIBLE AMERICA

Luzena Stanley Wilson experienced the horror of a burn-out at firsthand. She had lived in the small California community of Nevada City for only eighteen months when fire changed her life:

> Some careless hand had set fire to a pile of pine shavings lying at the side of a house in course of construction, and while we slept, unconscious of danger, the flames caught and spread, and in a short half hour the whole town was in a blaze. We were roused from sleep by the cry of "Fire, fire" and the clang of bells. Snatching each a garment, we hurried out through blinding smoke and darting flames, not daring even to make an effort to collect our effects. There were no means for stopping such a conflagration. Bells clanged and gongs sounded, but all to no purpose save to wake the sleeping people, for neither engines nor firemen were at hand. So we stood with bated breath, and watched the fiery monster crush in his great red jaws the homes we had failed to build. The tinder-like pine houses ignited with a spark, and the fire raged and roared over the fated town. The red glare fell far back into the pine woods and lighted them like day; it wrapped the moving human creatures in a fiendish glow, and cast their giant shadows far along the ground. The fire howled and moaned like a giant in an agony of pain, and the buildings crashed and fell as if he were striking them down in his writhings. When the slow dawn broke, and the sun came riding up so calm and smiling, he looked down upon a smouldering bed of ashes; and in place of the cheerful, happy faces, which were wont to greet his appearance in the busy rushing town of yesterday his beams lighted sad countenances, reflecting the utter ruin of their fortunes. The eight thousand inhabitants were homeless, for in the principal part of the town every house was swept away; and most of them were penniless as well as homeless.[14]

As Wilson probably knew, California's fire record was notoriously bad. According to one mid-century estimate, millions of dollars were lost there every week in "overwhelming conflagrations."[15] During the 1850s, the towns of Grass Valley, Columbia, and Sacramento were virtually wiped out by fire, and San Francisco, where blazes had wreaked havoc since the start of the Gold Rush, suffered a huge conflagration that destroyed more than three-quarters of the city.[16] As the state became more settled, condi-

California genre painter Christian C. Nahl and his brother painted this dramatic scene, *Fire in San Francisco Bay, July 24, 1853*. Red-shirted firemen used boats to support heavy water hoses while fighting the blaze. Courtesy of Fireman's Fund Insurance Company and Montgomery Gallery, San Francisco.

tions improved somewhat; but as late as 1905, when insurance inspectors toured San Francisco, they discovered serious fire hazards throughout the city. "In fact, San Francisco has violated all underwriting traditions and precedents by not burning up," the incredulous engineers reported. "That it has not done so is largely due to the vigilance of the fire department, which cannot be relied upon indefinitely to stave off the inevitable." Indeed, six months later, a violent earthquake touched off the citywide conflagration that has come to be known as the worst tragedy ever suffered by an American city.[17]

The situation on the East Coast was not much better. As early as 1830, New York City had racked up the unenviable fire record of three or four outbreaks per day; by 1850, property losses were thought to approach thousands of dollars daily.[18] Fires were so commonplace during the antebellum years that the city's firefighters became widely celebrated as tourist attractions. Shortly after his arrival in New York City in 1827, Basil Hall, a British naval captain, recorded a disappointing episode in his diary.

So much had been said to me of the activity and skill of the New York firemen that I was anxious to see them in actual operation; and accordingly, having dressed myself quickly, I ran down stairs. Before I reached the outer-door, however, the noise had wellnigh ceased; the engines were trundling slowly back again, and the people grumbling, not without reason, at having been dragged out of bed to no purpose. Of this number I certainly was one, but more from what I had lost seeing than from any other cause.[19]

Happily for Hall, he had another opportunity to chase fire engines that very night. This is hardly surprising, considering the city's vulnerability to flame. Between 1835 and 1855, New Yorkers reportedly suffered property losses of more than thirty million dollars from major fires alone.[20] During the second half of the century the fire department continued to be challenged—and generally defeated—by major conflagrations at such landmarks as the Crystal Palace (1858), the American Museum (1865), and the Brooklyn Theatre (1876). Though less dramatic, countless small blazes kept the fire alarms ringing and occasioned great financial hardship for individual householders and entrepreneurs during the same period. To this day, the New York City fire department is the busiest in the country.

California and New York were like giant andirons flanking an entire nation that threatened to go up in smoke. From about 1800 to 1858, Americans lost an estimated $255 million worth of property to fire.[21] Annual figures varied considerably, but the nation's insurance underwriters periodically came up with documented tallies of the destruction: a total national loss of $29 million in 1864, for instance, as contrasted to $36 million in 1880. In particularly disastrous years—1871 comes instantly to mind—the destruction of a single blaze far exceeded such composite figures. And in some years the national aggregate climbed to unusually high levels. In 1889, for instance, major fires in New York City, Seattle, Spokane, Lynn, and Boston contributed to a nationwide loss of approximately $123 million.[22]

Less sweeping statistics help to fill in the dismal details of America's fire history. Although the devastation can be measured in many ways, almost every conventional yardstick reveals a country plagued by what one early twentieth-century publication called "Enormous Fire Losses" and yet another dubbed our disgraceful national "extravagance."[23]

Not surprisingly, the most common approach toward assessing the damage concentrates on dramatic, citywide conflagrations. Taken as a group—and there certainly were enough of them to form a group—these episodes show a fairly steady evolution toward ever more destructive and more costly incidents. Thus the New York Merchant's Exchange fire of 1835, in which some 530 buildings were leveled, was soon eclipsed by the 1845

"The Burning of Chicago," a lithograph issued shortly after the conflagration, captured the chaos of the scene. Peters Collection, National Museum of American History, Smithsonian Institution Photo 60451-B.

Pittsburgh fire, which consumed 1,100 buildings. In 1851 the dubious honor was transferred to San Francisco in commemoration of the early May blaze that destroyed 2,500 buildings and $17 million worth of property over at least three-quarters of the city's acreage. Not long thereafter, the torch was passed to Portland, Maine, where a firecracker touched off to celebrate the Fourth of July in 1866 ignited flammable materials stored in a boat shop and started the infamous conflagration that destroyed half a mile of the business district and left 10,000 people homeless.

With the great Chicago Fire of 1871, this statistical game reached epic proportions. The stage for Chicago's disaster was set when an excessively dry autumn caused tinderbox conditions throughout the region. On the fateful night of October 8, a cow reportedly kicked over a lantern in the O'Leary family barn and touched off a blaze that quickly raged out of control. Fanned by a capricious wind and fueled by block after block of dry wooden structures, the fire swept across the city. Lacking adequate water supplies and assaulted by the fierce heat and wind, firemen already exhausted by weeks of firefighting could do little to retard the destruction. By

THE GREAT FIRE AT BOSTON,
NOVEMBER 5th & 10th 1872.

The Fire began on Saturday evening, and raged for 15 hours, destroying over Sixty Acres of Buildings, among which were whole blocks of the finest Granite Stores on the continent, and property estimated at nearly $ 100,000,000.

The Boston fire of 1872 was a popular subject for lithographers of the period. Peters Collection, National Museum of American History, Smithsonian Institution Photo 60451-E.

one estimate, the fire devoured sixty-five acres of cityscape every hour until, about twenty-seven hours later, a rainstorm finally extinguished the flames. Property losses amounted to about $200 million, with about 100,000 people (one-third of the city's population) left homeless. Associated losses included eight bridges, 121 miles of sidewalks, and more than 16,000 buildings.

Subsequent "great" fires had different features but similar disaster quotients. The Boston fire of November 9–10, 1872, though largely confined to the commercial district, destroyed 776 buildings across sixty-five acres for a financial loss of about $75 million. Just after the turn of the century, Baltimore suffered an even more destructive fire in its business district. Downed electrical lines added to the fury of the flames, which burned nonstop for more than twenty-four hours. At final count, the destruction encompassed almost 140 acres of property worth $150 million.[24]

These blazes were just warm-up exercises for the "big one," the earthquake-induced fire that leveled San Francisco in 1906. Still a record holder, this fire caused such disruption to normal existence that armed guards were called in to help control the panicked population. Jack London, one of many celebrities to witness the disaster, described the scene vividly for *Collier's Weekly*:

Within an hour after the earthquake shock the smoke of San Francisco's burning was a lurid tower visible a hundred miles away. And for three days and nights this lurid tower swayed in the sky, reddening the sun, darkening the sky, and filling the land with smoke. . . . Dynamite was lavishly used, and many of San Francisco's proudest structures were crumbled by man himself into ruins, but there was no withstanding the onrush of the flames. Time and again successful stands were made by the fire fighters, and every time the flames flanked around on either side, or came up from the rear, and turned to defeat the hard-won victory.[25]

When the air finally cleared, the damage was almost unbelievable. In seventy-four hours of fury, the fire had burned across 4.7 square miles, destroyed more than 28,000 buildings, and rendered 225,000 people homeless. It was readily conceded that the destruction amounted to about $350 million, although some analysts argued that the total financial loss more realistically approached half a billion dollars.[26] To one shocked survivor, the scene was "the most awful sight I ever saw. . . . It looked as if Hell had opened its jaws."[27]

Such underworld imagery was common rhetoric at the time and was just as appropriate for small urban fires as for big ones. While the great conflagrations grabbed national headlines, less dramatic fires consumed small towns and communities with equally relentless ferocity. Cumberland, Maryland, was completely destroyed by fire in 1833. Similar disasters struck Providence, Ohio, in 1846 and Columbus, Iowa, in 1847. Western mining towns, with their flimsy construction and insufficient water resources, were especially vulnerable to fire. Nevada City, California, was virtually destroyed at mid-century, and Tombstone, Arizona, burned to the ground twice during its heyday. As late as 1912, Frog Point, Michigan, suffered a devastating blaze that, in the words of its local editor, "practically wiped the town off the map."[28]

Meanwhile, America's urban fire losses were augmented by a variety of wildland fires that swirled menacingly around the outskirts of civilization and periodically inflicted their own form of devastation. Prairie fires were the most notorious of these threats because their appearance was so regular and because they endangered so many homesteads across the plains. Once ignited—whether by careless campers, lightning, arsonists, or farmers themselves—these blazes continued for many weeks each autumn, devouring the prairie's brittle dry grasses and anything else that happened to be

13234—Militiamen on Guard, Burned District, Baltimore, Md.,
U. S. A.

Opposite, top: Ruins after disastrous fire in Omaha, Nebraska. Stereograph by Currier, Omaha.
Opposite, bottom: "Militiamen on Guard, Burned District, Baltimore, Md." Stereograph by
Keystone View Company, 1904. *Above, top:* "City Hall, San Francisco After Earthquake and
Fire Disaster." Stereograph by Keystone View Company. *Above, bottom:* Remains of the
Franklin School, destroyed in the Chicago Fire of October 9, 1871. Stereograph by
Lovejoy and Foster, Chicago.

in the way. This powerful "sea of fire" was "a strange and terrible sight" to immigrant pioneer Gro Svendsen. "In dry weather with a strong wind," he claimed, "the fire will race faster than the speediest horse."[29]

Forest fires unfolded in the same menacing way. In fact, the costliest fire disaster in United States history in terms of lost lives was the forest fire that decimated Peshtigo, Wisconsin, and a number of surrounding towns on October 8, 1871. Coincident with the Chicago fire, this ferocious blaze swept across 1.25 million acres of dry timberland in a tumultuous whirlwind of fireballs, ashes, and thick black smoke. The word *firestorm* originated with this incident.

Disastrous multiacreage fires had long blackened the American landscape. During the autumn of 1825, for instance, 2 million acres had burned along the Miramichi River in New Brunswick, along with 800,000 additional acres in Piscataquis County, Maine. Aided and abetted by careless lumbering practices and spark-spewing locomotives, fire history continued to repeat itself. The Great Lakes region, in particular, suffered massive destruction by woodland fires between 1870 and 1918. Wherever settlers went, however, the problem followed. Probably the worst such incident on record was the 1910 holocaust in which five million acres of national forest land burned in the far western part of the country.[30]

Thus, again and again, in town and country alike, Americans encountered the disturbing scent of burnt money. And yet, although an overview of the "great fires" helps to establish the dimensions of the problem, the preoccupation with statistics almost trivializes the severity of the destruction. Consider, for instance, the extraordinary losses resulting from fires in the nation's libraries. Notable examples include the fires at the Baltimore Lyceum in 1835, the Library of Congress (which housed Thomas Jefferson's personal library) in 1851, and the Chicago Historical Society in 1871 (which then housed the Emancipation Proclamation).

Less dramatic but equally injurious were the innumerable courthouse fires that routinely destroyed vital personal and property records throughout the country. In the short run, these losses wreaked havoc on citizens trying to rebuild their lives after disastrous municipal fires had destroyed their homes and businesses. (It took more than half a century to sort out the legal tangle created by the incineration of San Francisco's paperwork in 1906.) In the long run, the widespread destruction of the documentary underpinnings of American life has left history the loser.

Prime examples of the nation's ingenuity and artistry have been consumed by fire as well. The great Western painter William Keith lost more than two thousand paintings when his San Francisco studio burned in 1906. The nation's tinkerers lost as many as 75,000 examples of their own artistic endeavors when the Patent Office and its collection of models burned in 1877. Architects working in wood were particularly vulnerable. During a celebrated career, Lewis Wernwag designed several dozen bridges, including "The Colossus," a magnificent 340-foot wooden arch built over the Schuylkill River in Philadelphia in 1812. For twenty-six years the bridge attracted visitors and brought publicity to the city; and then, on the night of September 1, 1838, the bridge was consumed by fire.[31]

Builders risked disappointment every time they practiced their trade. The Chestnut Street Theater, the best house in Philadelphia, was destroyed by fire in 1820. Tremont House, the best hotel in Chicago, burned in 1839 and then again ten years later. Devastating fires wiped out Lexington's Transylvania University in 1828, Milwaukee's fabulous new music hall in 1852, and Boston's famed Trinity Church in 1872. Thomas Jefferson's Rotunda graced the University of Virginia campus until a fire in 1895 ravished the handsome interior of the building. The Wayside Inn, built in 1690 and immortalized by Longfellow, survived much longer, but in 1955 it, too, was destroyed by fire.[32]

These last two structures had such historical significance that they were quickly reconstructed by citizen groups determined to preserve the past. For individuals, however, the rebuilding process was not always so swift or simple. Although victims of serious fires in any era suffer emotional dis-

William Coventry Wall (1810–1886), *Pittsburgh After the Fire of 1845 from Boyd's Hill* (oil painting, 1845). Wall's painting was also distributed in the form of lithographs. Courtesy of the Carnegie Museum of Art, Pittsburgh; gift of the estate of Mary O'Hara Darlington, 1925.

tress, the increased risk of fire in earlier times adds poignancy to the stories of the survivors. William Cumming Peters, Stephen Foster's publisher, suffered two fires in quick succession at his warehouse. The great showman P. T. Barnum experienced three devastating losses in his lifetime: his beloved home, Iranistan, burned to the ground in 1857, and his world-famous American Museum succumbed to the ravages of the flame twice. For Cyrus McCormick, the irascible inventor of the reaper, the problem of fire was vexing in the extreme. He lost his luggage in a railway fire in 1862; then, nine years later, while he was still seeking recompense for that loss, his entire factory burned up in the Chicago fire of 1871.[33]

"My pleasant things in ashes lie, and them behold no more shall I," lamented poet Anne Bradstreet after a fire destroyed her house in 1666.[34] Two centuries later, a western woman experienced the same profound sense of loss when a fire destroyed her childhood home. "I wept and wept over my doll, the grand riding hat and feather, and the beautiful treasures in the cellar," she wrote. "They were all the world to me then."[35] In such a highly flammable world, it must have seemed to many that the only enduring reality was the destructive power of the flame.

SMOKE-FILLED AMERICA

"There is no fire without some smoke," observed a proverb current in the nineteenth century. And, as everyone knew, smoke—like fire—caused a wide range of problems.

Despite its routine appearance during combustion, smoke is not, properly speaking, a product of combustion. According to one chemistry textbook popular in the nineteenth century, it is simply fuel "which escapes combustion, and which, by its condensation, forms the soot in our chimneys."[36] This explanation was probably only marginally less mysterious to the average American than the "smoke-lore" of the day, which claimed, among other things, that smoke was a ghostly messenger to the spirit world. But at least the textbook referred to something tangible: soot. This, along with the other evil side effects of smoke, was something that everyone who used the open flame could understand.

Domestic Smoke

Nowhere were the evils of smoke more annoying than in the home. Colonial lighting devices such as candlewood torches, rushlights dipped in grease, and Betty lamps fueled by animal fat or fish oil were notoriously smoky. So were tallow candles, particularly those made from hog's lard, which, one expert noted, "occasions a thick black smoke."[37] The coal oil and kerosene lamps that supplanted these rustic illuminants introduced far less smoke into the air, but they had the annoying habit of accumulating black soot on their glass chimneys. Householders maintained vast quantities of "lamp rags" for the perpetual and thoroughly distasteful task of cleaning such lamps. Edith Stratton Kitt, one of countless pioneers to suffer lamp duty, recalled her childhood experiences in Arizona: "In the mornings we could do pretty much as we liked, but we had a few regular chores. What we disliked most was to fill the coal-oil lamps, trim their wicks, and clean their sooty chimneys."[38] In wealthier homes, servants were commonly assigned the grimy task.

Fires for heating and cooking were even more problematic than those kindled for illumination. This was especially true for those individuals forced to rely on open-air campfires in which wind and wet wood often conspired to produce excessively smoky conditions. "It is very trying on the patience to cook and bake on a little green wood fire with the smoke blowing in your eyes so as to blind you," recalled a young woman journeying west in 1852.[39] Yet even homemakers blessed with masonry fireplaces suffered from chronic smoke.

Part of the problem, as Benjamin Franklin pointed out, was the injudicious design of the majority of American fireplaces and chimneys. In a provocative letter published posthumously in 1793 as *Observations on Smoky Chimneys, Their Causes and Cure*, Franklin pointed out nine major design flaws, including excessively large fireplace openings, excessively short chimney shafts, and the lack of adequate air circulation in the room containing the fireplace. Other well-known personages also viewed the smoke problem as a design problem. Benjamin Thompson's innovations included narrowing the throat of the fireplace and installing a "smoke shelf" to prevent downward currents of air from blowing smoke into the room. Charles Willson Peale's contribution was the "smoke eater," a device that forced smoke back through the firebox in order to remove the soot. Countless other tinkerers immortalized their own pet solutions for smoky chimneys at the U.S. Patent Office.

Yet, the domestic smoke problem also arose from variables unrelated to construction techniques. Smoke could be increased by such factors as the type and condition of the fuel (wet or green wood was bound to smoke), the weather (excessive cold or wind affected the draught of the chimney), and the level of cleanliness of the hearth and stack (soot accumulated on the interior walls of the chimney was fuel waiting to smolder). Fixing one problem did not necessarily eliminate the contaminant. The introduction of cast-iron cooking and heating stoves helped to control smoke in many cases; but even these devices, with their enclosed fireboxes and elaborate damper systems, failed to clear the air completely. Smoking chimneys were especially common when several stoves were connected to a single flue, but there were more fundamental difficulties as well. As one expert put it, "Every one knows that smoke and gas frequently escape from the joints of the common stove when the door is open, and also when the damper in the smoke pipe is shut."[40]

Everyone must have known, moreover, the many unpleasant side effects of this insidious form of pollution. "Damaged furniture, sore eyes, and skins almost smoked to bacon," were the key problems in Benjamin Franklin's view.[41] Dirty houses, food tainted with the taste of smoke, and damaged clothing were other common complaints. For women concerned with their youthful appearance, smoke was a menace. "I have cooked so much out in the sun and smoke that I hardly know who I am . . . when I look into the little looking glass," complained Miriam Davis, who settled in Kansas in 1855.[42] Other women noted the deleterious effects of smoke on their health and general attitude toward life. Carbon monoxide, which leaked imperceptibly out of almost all fuel-efficient stoves, caused headaches and fatigue, and smoke exacerbated any number of respiratory complaints.

Joseph Scarlett, an enterprising chimney sweep knew these problems well. "The life of e'en a king will be uneasy [if] wrapped in smoky pall," he claimed in an eye-catching advertising brochure. Although his chimney-cleaning service could hardly eliminate smoky rooms entirely, Scarlett unquestionably succeeded in defining the scope of the problem:

> A smoky house, a crabbed, scolding wife,
> Are little evils, and of old complaint;
> They make poor souls a-weary of this life,
> Provoke the curses of a very saint.[43]

Smoke in the Community

One of America's lesser-known tourist attractions in the early nineteenth century was the "Italianate" clarity of its air. As John M. Duncan discovered during his visit in 1818–1819, New Englanders enjoyed "a purer atmosphere" than Old Englanders, largely because "the smoke of coal fires is unknown."[44] Duncan saw the country at the right time. As the population grew and fuels changed, the picture clouded. By the second half of the nineteenth century smoke had become a problem in many places, and in heavily industrial centers it presented a serious threat to buildings, vegetation, and the health of the inhabitants.

The exact nature of the problem depended on the composition of the fuel and the mode of combustion. Bituminous coal, which was discovered in extensive deposits in western Pennsylvania and throughout the Mid-

Active smokestacks, those symbols of industrial progress, could also present problems for the inhabitants. This view shows that Utica, New York, was smoking moderately by mid-century. Peters Collection, National Museum of American History, Smithsonian Institution Photo 55203.

west and West, was especially troublesome. If burned before coking, this so-called soft coal was highly volatile and hence smoky—so smoky, in fact, that the U.S. Navy was forced to discontinue its use because the fleet was sending out unintentional smoke signals to the enemy. Additionally, bituminous coal had the disadvantage of containing a high percentage of sulfur, which, when released into the atmosphere during combustion, was transformed into corrosive and highly destructive sulfuric acid. This problem-ridden fuel represented only one bad choice among many, however. Complete combustion was a virtual impossibility except in the most carefully controlled laboratory situations. Elsewhere, whenever anything was burned, undesirable substances invariably escaped into the atmosphere. People relied heavily on air currents to cleanse their surroundings, but this was not always enough.

Passengers on the nation's railroads became aware of this depressing fact early in the nineteenth century. One did not require Henry David Thoreau's powers of observation to see that the typical locomotive of the 1840s and 1850s was like a powerful dragon "breathing fire and smoke from his nostrils."[45] A short ride on a local line would have validated the simile in a matter of minutes. The nation's inventors and tinkerers confronted the problem head-on and designed countless dust catchers, spark arresters, ventilators, and smokestack grates to cut down on smoke and cinders. In one particularly innovative plan, a Massachusetts technician sought to purify noxious locomotive smoke and gases by passing them through water. "Travelers on railroads would be very grateful if some such plan were adopted," declared one account of this novel but impractical scheme.[46] Travelers would undoubtedly have been grateful for any improvements, and a number of inventions were adopted. Nevertheless, train travel continued to be a gritty proposition throughout the century. As the English statesman James Bryce discovered during his many visits to the United States, it was a cherished American custom to reserve the end railroad car—"that farthest removed from the smoke of the locomotive"—for the ladies.[47]

The ladies fared less well on the streets, especially in Pittsburgh, where soot and cinders from local iron and steel mills rained down upon them. It's "Come to Blows!" reported *Scientific American*, playfully suggesting that women so afflicted should blow, not rub, the city's debris from their garments.[48] But what was laughable on one level was also cause for serious concern. As early as 1816, traveler David Thomas was able to identify Pittsburgh by its smokiness when still a half mile away: "Dark dense smoke was rising from many parts, and a hovering cloud of this vapour, obscuring the prospect, rendered it singularly gloomy."[49] Close up, the city presented an even more somber appearance, particularly as the century wore on. "The buildings, whatever their original material and color, are smoked to a uniform, dirty drab," wrote Willard Glazier in 1885. "The smoke sinks, and mingling with the moisture in the air, becomes of a consistency which may almost be felt as well as seen."[50]

Other communities had similar, though less notorious, problems with smoke. St. Louis suffered so much from air pollutants that by the middle of the century the city government passed an ordinance regulating the height of chimneys. Apparently these measures were insufficient. In 1864, in one of the earliest smoke cases to come to trial, a local man sued and won fifty dollars for smoke damages he had sustained. Though less populous, western mining communities such as Black Hawk, Colorado, also faced the problem of chronically darkened skies. One visitor in 1877 found the situation there alarming. "Immense smelting works fill the atmosphere with coal dust and darkness," he wrote in a narrative that appeared in *Engineering and Mining Journal*. "One mill at our right is sending forth volumes of blackness from *seventeen* huge smoke stacks."

In many localities such problems were exacerbated by the presence of toxic vapors in the smoke. Smelter smoke containing arsenic, bismuth, and lead engulfed the citizens of Leadville, Colorado, in a serious health crisis during the 1870s. In Butte, Montana, the copper capital of the country, the problem was even more severe. Frequently blanketed by a "complete envelopment" of smoke and fumes, the city suffered not only from daytime darkness but also from health problems caused by air-born soot laced with arsenic and sulfur. Emma Abbott, a famous opera singer, gave one concert in the town and ran for her life.[51]

For residents of smoke-filled communities the best course of action was antismoke legislation, which emerged piecemeal during the last two decades of the century. Chicago passed the first serious smoke ordinance in the country in 1881. It stipulated that "the emission of dense smoke from the smokestack of any boat or locomotive or from any chimney anywhere within the city shall be . . . a public nuisance." Penalties ranged from five to fifty dollars. Shortly thereafter, Cincinnati, Pittsburgh, Cleveland, and St. Louis put similar laws on their books. The situation in mining districts was generally complicated by the overwhelming corporate presence in town affairs, but the outraged citizens of Butte finally managed in 1891 to compel smelter operators to find a way to cut down on smoke or face a two-hundred-dollar fine.

These and other measures were hardly solutions to the problem. Rather, they served as guidelines for future policy. "Air pollution" is a twentieth-century term for one of the major challenges of twentieth-century government. Yet the stage was set—indeed, the problem was introduced—in the previous century. J. W. Goldthwait's diary of a train trip west during the summer of 1902 serves as a link between the two eras. Consider these observations by a man on the edge of the new century:

> June 28: [after a train journey] . . . washed off the cinders & smoke before going to bed.
> July 2: Pittsburgh's striking features are industry and smoke, in equally large proportions. Can tell the relative age of the buildings by the degree of dinginess.
> [In Chicago] . . . went to the Montgomery Ward & Co. new skyscraper, one block away. . . . Could look directly down into the big city & see way off on the lake; but smoke obscured much.
> [Back on train:] Sat for a while on observation platform at rear of train, with H & Cobb. Cinders too thick for comfort, so didn't stay there long.[52]

Even a century ago, there was a lot of work to be done.

Burns, Scalds, and Other Agonies

Medical men of the nineteenth century characterized burns in various ways. Jacob Bigelow, writing in the *New England Journal of Medicine and Surgery* in 1812, offered colleagues the following impassive observation: "The application of substances to the human fibre, which are heated beyond a certain temperature, is followed by the phenomena of pain and inflammation."[53] John C. Gunn, the author of an immensely popular treatise on domestic medicine in the 1830s, took a different tack: "Because we all know well what scalds and burns are . . . I shall not attempt to describe them."[54] That said, Gunn forged ahead with his recommended remedies.

Daniel Drake, a naturalist and physician from Cincinnati, had yet another approach. Having suffered a serious burn himself, he was able to convey, in all its horror, the agonizing reality of a serious burn. "It would be difficult to set forth the variety of physical and moral sufferings, which were attendant on the protracted state of nervous or constitutional irritation which [this burn] generated," he wrote in his own case study of 1831. "In short every sensation both of mind and body was unpleasant if not painful; and this continued to be the case for at least five weeks." Drake's hands, which had sustained the brunt of the injury, were the source of almost unbearable suffering. Long after the accident, he could still recall the excruciating effects of the burn:

> The granulations became affected with a distressing neuralgia, which was violent, for six or eight months and has not yet, at the end of nearly two years, entirely ceased. It would be difficult to characterize the different kinds of pain which the patient endured for the first half of this period. Sometimes it was dull and aching but oftener acute and suddenly radiated from a point through the whole eschar. At other times, it resembled the sensation produced in the extremities of a nerve, . . . but more commonly the patient could only compare it with the imaginary effect of millions of fine, red hot needles, running in all directions through the new flesh.[55]

Drake sustained his injury while trying to save his sister-in-law from a bed that had been set ablaze by an untended candle. It was a common accident at that time but by no means the only source of domestic burns. Women who stood at the fireside in order to cook were in constant peril from sparks and flames as well as from the scalding heat of overturned kettles. Their long and voluminous dresses only compounded the problem. Children were also at extreme risk. As an infant, Lucy Larcom was dropped into the hot coals by a half-blind aunt who was cuddling the new baby in the chimney corner. Larcom escaped this mishap uninjured, but plenty of other children were seriously burned as they played by the fireside or, more tragically, when they tripped and fell into the flames. Drunken adults commonly suffered a similar fate, and even those who were sober could slip unconsciously into red-hot embers after falling asleep in the comforting glow of the hearth. Gunn reports the horrifying case of a man who suffered a fit and fell into the large fire by which he was sitting after his family had

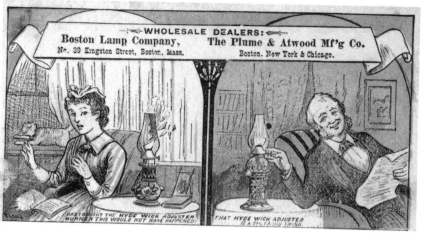

-=WHOLESALE DEALERS:=-
Boston Lamp Company,   The Plume & Atwood Mf'g Co.
No. 39 Kingston Street, Boston, Mass.   Boston, New York & Chicago.

The danger of exploding lamps provided a dramatic illustration for this
lamp company trade card, ca. 1870.

gone to bed. By the time he was discovered, the poor man was "roasted,"
his ribs exposed and the motion of his intestines purportedly visible. Mi-
raculously, the man did not die, although he probably wished that he
had.[56]

With the introduction of stoves and lamps, household dangers were
reduced somewhat, but the lighting and maintenance of such devices still
required hands-on manipulation of fire. Lamps fueled by various "burning
fluids" could—and did—explode with dire consequences. Wash day continued to be a nightmare, the source of painful skin irritations caused by
exposure to fire, boiling water, and steam.

Outside the home, the dangers were varied and widespread. The great
conflagrations often had huge death tolls: an estimated 250 in Chicago
(1871), 300 in the fire associated with the Johnstown Flood (1889),[57] and
more than 500 in San Francisco (1906). In the most deadly American fire
of all, the Peshtigo forest fire, an estimated 1,500 people lost their lives.

Yet, despite the publicity accorded these big-area tragedies, fires in
crowded, enclosed places were the real killers.[58] Theaters, with their continual use of the open flame amidst the highly combustible fabrics of the
stage world, had a terrible record on this score. When the Brooklyn Theatre burned during a performance in 1876, an estimated 300 patrons lost
their lives. This widely lamented tragedy was surpassed a mere quarter
century later when, in 1903, Chicago's Iroquois Theater burned and killed
602 people. Other venues could be almost as dangerous. A dormitory fire
at Howard College in Alabama, for instance, caused about ten students to

Many inventors hoped to save lives with their ingenious fire escapes. The 1857 extension ladder (left) and 1868 pulley system were illustrated by Benjamin Butterworth in *The Growth of Industrial Art* (1892).

jump to their deaths during the tragic nighttime blaze of October 15, 1854. The more famous factory fire at the Triangle Waist Company in New York in 1911 resulted in the deaths of 145 young women.

Most of the Triangle workers could have been saved had management not sealed the exits. Other laborers, however, were at risk all the time because of the nature of their work. Blacksmiths and ironworkers exposed themselves daily to the dangers of the open flame, and coal miners risked injury from exploding mine gases every time they descended into the shafts. Between 1855 and 1910, at least 3,000 miners perished in the two dozen or so major coal mine disasters that involved explosions or fires.[59] For farmers, the dangers were less severe, but they too risked injury every time they torched underbrush or operated steam-powered equipment.

As contemporary observers liked to point out, public transportation during the last century could be extremely hazardous to one's health. Steamboat explosions were distressingly common in the days before the Civil War. By one estimate, some 2,500 people died between 1817 and 1850 as a result of boiler explosions on Mississippi River packet boats.[60] As one horrified witness suggested in his journal, those were the lucky ones. Walter B. Foster, who was on the scene when the steamship *Edna of the Platte* exploded in 1842, describes the "screams and groans, curses and prayers" of the pitiful survivors of this terrible disaster. Those who had inhaled steam screeched in agony, and one poor woman who had scalds over most of her body begged Foster to remove the injured skin that dangled like a glove from her arm.[61]

THE BURNING COAL MINE AT WILKESBARRE, PENNSYLVANIA—STOPPAGE OF THE "FAN."—SKETCHED BY ALFRED J. EAMES—SEE PAGE 90.

Images and stories of burning coal mines, with the attendant drama of miners trapped deep underground and frantic rescue efforts, portrayed the everyday dangers in this profession. This woodcut appeared in the April 18, 1874, issue of *Frank Leslie's Illustrated Newspaper*.

Although the nation's railroads liked to emphasize their relative safety, train travelers often suffered from burns and scalds as well. Sparks emitted from the engine fires were a hazard from the very beginning, as a passenger on the *DeWitt Clinton*'s 1831 maiden voyage discovered: "As there were no coverings or awnings to protect the deck-passengers upon the tops of the cars from the sun, the smoke, and the sparks, and as it was the hot season of the year, the combustible nature of their garments, summer coats, straw hats and umbrellas, soon became apparent, and a ludicrous scene was enacted."[62] It was far from ludicrous, however, when sparks caused serious injury or when locomotive fires ignited entire trains in the wake of collisions and derailments. Passenger car heating stoves were a terrible menace, causing superficial burns to passengers jostled against the hot metal during routine runs and setting the cars ablaze during serious accidents. A great public outcry against these stoves gained momentum

TERRIBLE CONFLAGRATION AND DESTRUCTION OF THE STEAMBOAT "NEW JERSEY,"
On the Delaware River, above Smith's Island, on the Night of March 15th, between 8 and 9 o'clock, in which dreadful calamity over 50 Lives are supposed to have been lost.

Steamboat fires claimed thousands of lives and provided dramatic images for lithographers. Peters Collection, National Museum of American History, Smithsonian Institution Photo 60451-D.

during the 1880s and led to the adoption of safer steam heating devices on most of the nation's lines. Strangely, however, the manufacturers of the new steam heaters kept the fear of fire alive through their advertising. "No Fires in the cars, No one Roasted Alive," boasted the Martin Anti-Fire Car Heater Company of Dunkirk, New York, in what was supposed to be a reassuring testimonial.[63]

Faced with such risks, the country's most conservative travelers may well have preferred to walk. In the days of the open flame, however, even that course was fraught with danger, as the example of a couple from Delaware makes clear. Strolling on the quiet streets of Laurel one autumn day in 1899, Charles Adams dropped a lighted match in a pile of leaves, which immediately caught fire and ignited his wife's skirts. "Her screams attracted his attention when he was looking for nuts," reported the local *Morning News*. Luckily, Adams "managed to peel part of her dress off and save her."[64]

In short, anyone who used fire risked injury from it—which is simply another way of saying that everyone in the nation was in danger. Not

Sparks, cinders, and exploding boilers were among the dangers that confronted early train passengers. The "barrier car," loaded with hay bales, provided some protection from such risks. From William H. Brown, *History of the First Locomotives in America* (1877).

"The modern altar of sacrifice—the devouring car stove," a woodcut from *Harper's Weekly* in 1887, decried the dangers of lighted stoves in railroad passenger cars.

surprisingly, fire safety was a matter of serious concern and was discussed repeatedly in private conversation and in print. Andrew Brown of Philadelphia took the spiritual approach, advising his children to "be good" in order to avoid becoming fire victims during the winter of 1797.[65] Other suggestions were much more specific. Catharine Beecher and Harriet Beecher Stowe implored readers of The American Woman's Home (1869) to avoid reading in bed lest they fall asleep and accidently set the bed on fire.[66] Alexander V. Hamilton, who was shocked by the careless fire habits of American workers and homemakers, advised in his Household Cyclopaedia (1875) that matches be "cautiously used" and all lights extinguished at night.[67] Numerous precautions were advocated expressly for the protection of youngsters, whose innate curiosity rendered them particularly vulnerable to burns and scalds. The installation of sturdy fireplace fenders or screens and the adoption of woollen clothing (which was considered less combustible than cotton or linen) were among the most commonly publicized strategies. Busy pioneer women are also known to have resorted to the more basic expedient of tying their children in bed in order to prevent toddlers from wandering into fires.[68] For extinguishing clothing that had caught fire, the "stop-drop-and-roll" procedure had become standard by mid-century. Schoolchildren in 1900 were taught in their primers to smother the flames downward in order to protect the face and hair of the victim. Cookbooks often carried similar advice.

Though philosophically sound, these and other fire-safety techniques were based more on hope than on experience. In a society that depended on the open flame, burns and scalds were all but inevitable. For physicians and homemakers alike, therefore, the key question was not how to prevent burns but how to heal them.

Folklore often offers a single magic cure for burns, whereas real-life treatment was incredibly diverse. Hot water, cold water, nitrate of silver, sulfate of iron, bicarbonate of soda, white paint, and sulfate of zinc are just some of the remedies advocated in medical journals of the nineteenth century. If those substances failed, carbolic acid and linseed oil, subnitrate of bismuth, spirit of turpentine, electricity, treacle, and Irish potatoes were other recommended options. According to Lydia Maria Child, cotton wool and oil were unquestionably "the best things for a burn." By the same token, she also knew of a case where the "constant application of brandy, vinegar, and water" was effective, particularly if accompanied by a good swallow of paregoric.[69]

The lack of agreement on a cure stems from the fact that every burn, like every burning building, was unique. The cause of the burn, the area and the depth of the injury, and the general physical condition of the patient were all factors to be considered before treatment could begin. To simplify the situation somewhat, physicians throughout the nineteenth century divided burns into three now-familiar types, each requiring different levels of treatment. Therapy for classes one and two (characterized by a reddening of the skin and by blistering, respectively) generally entailed application of ointments or lotions, puncturing the blisters, and careful bandaging of the wound. For class three burns, which involved much more severe skin damage, physicians had to choose from a wide range of topical and constitutional treatments. But such classifications were useful only up to a point. Burns could be very complex, often entailing serious side effects such as shock, tetanus, gangrene, and debilitating scarring. Furthermore, an extensive first-degree burn could be just as fatal as a smaller injury of the third degree. The medical literature on burns and scalds during the last century is, consequently, a vast collection of individual case studies. Not infrequently, a stated therapy worked in one case and not another.[70]

The general public was not aware of these professional publications and did not need to be. By and large, they favored an approach to burn therapy that excluded physicians from all but the most severe cases. For many Americans, particularly during the decades before the Civil War, when population density was low and distrust of doctors was high, lay practitioners provided a happy alternative to professional medical care. Word-of-mouth was often sufficient to establish the reputation of these healers (one woman tells of a man hiking six miles because he had heard good things about her burn-curing methods), but advertising was not uncommon. As early as 1808 Judith Corey informed readers of the *New England Palladium* that she not only served as a midwife but also cured burns.[71] Many citizens rejected consultations altogether and simply turned to their household manuals for burn remedies that they could initiate on their own. The best-known of these treatises were William Buchan's phenomenally popular *Domestic Medicine* (first published in the United States in 1771 and reprinted many times thereafter) and John C. Gunn's *Domestic Medicine*, which appeared in 1830 and remained popular for decades. Countless others were produced throughout the century. As late as 1897, long after doctors had improved their reputations nationwide, the author of *The Cottage Physician* made a strong case for "domestic medicine" in the treatment of burns: "So constantly are these painful accidents occurring, and so frequently does it happen that the care of a medical man cannot be obtained for them, that it behooves all heads of families to make themselves acquainted with the best remedial measures."[72]

There was wisdom in this approach. Treatment could begin as soon as the injury occurred. Furthermore, because the manuals tended to emphasize the use of common household commodities—"remedies as are immediately at hand, or are easily obtained," as Gunn put it—the procedures were straightforward and devoid of mystery. With the proliferation of patent medicines during the last quarter of the century, home treatments became even more streamlined and economical. Finally, there was a certain poetic justice in household medicine. Throughout the nineteenth century, people built their own fires, fueled their own fires, and fought their own fires. In such a society, it seems only appropriate that they should also cure their own fire-related injuries.

Frequently, home medicine worked. In a detailed letter published in Dr. Samuel Latham Mitchill's *Medical Repository* in 1820, a housewife name Hannah Barnard described her many successes using an ointment of Burgundy pitch, beeswax, and oil. In one of her cases, a young woman had suffered a scald that raised a blister on about one-third of her arm. "I immediately applied the plaster, and bound it up close," writes Barnard. "It immediately gave her complete relief from any further suffering." Barnard allowed the bandage to remain in place undisturbed for four or five days so that recuperation could take place. Happily, a week after the accident, the scald had "compleatly healed, and no inflammation ever appeared in it."[73]

There were, however, less successful stories to tell. Sometimes would-be physicians failed to cure their patients because they did too little. Sometimes they failed because they meddled too much.[74] Often the injuries were simply too severe to respond to any known treatments, and, after suffering agonizing pain, the victim succumbed.

Clara Guthrie, Woody Guthrie's sister, was one such victim. The family's coal oil stove exploded one day and set the girl's dress on fire. "Some of the neighbors chased her around the house," one account of the incident reports. "But they couldn't stop her, and the wind just made her dress burn that much faster."[75] Clara died a short time later. Her story is remembered because of the fame of her brother, but her mishap was hardly unique.

Bad Servant

The destructive side of fire was so commonplace and so easily set into motion that it was frequently exploited by antisocial individuals or groups seeking to inflict harm. Not infrequently, the damage was directed toward themselves. Though it is impossible to know how many people who "fell" into their blazing fireplaces actually jumped in, contemporary newspapers leave little doubt that suicide was occasionally a factor. Obituaries offer additional grisly evidence of crazed or desperately unhappy people who set fire to their own clothes or who deliberately torched their houses and then refused to flee. One day in February 1898, an old woman from rural Wisconsin reportedly doused herself with kerosene and lit a match. "She died in a horrible agony before help reached her," reported the local paper. "She was undoubtedly insane."[76]

The same might have been said about those who used the destructive side of fire to hurt others. The subject, though vast, can be divided into two distinct categories: fire-based punishments (those barbaric practices, such as branding and burning alive, that are now all but forgotten) and arson (which, unfortunately, is not).

Branding

Branding was employed throughout the colonies during the seventeenth and eighteenth centuries as one of many methods to deal with scofflaws. As with other corporal punishments (such as whipping and ducking), branding inflicted pain and humiliation on the wrongdoer. But branding, like ear cropping, had the added benefit of causing a permanent mark. A person was literally branded for life. Most colonies developed elaborate codes whereby a diagnostic letter was emblazoned onto a particular body part for a specific offense: B on the right hand for burglary, F on the forehead for forgery. In Maryland, where branding was unusually widespread, each county maintained a complete set of branding irons to punish crimes ranging from seditious libel (SL) and roguery (R) to thievery (T) and manslaughter (M). Even the humanitarian legal code created by the Quakers in Pennsylvania between 1682 and 1718 included branding in its later provisions, a surprising development considering the treatment Quakers had received in New England during the seventeenth century. Whipped and branded and driven from the region, Quakers risked having their tongues bored with red-hot irons if they returned to Puritan territory.

From time to time branding also played a part in the curious legal maneuver known as "benefit of Clergy." Originating in Britain as a means of protecting clerics, "pleading the clergy" eventually developed in the colonies into a legal loophole whereby almost anyone could escape severe penalties for some crimes. Instead of receiving a sentence of death or long imprisonment, the criminal was simply branded on the thumb. The formula did not work for all crimes or for all people—and, as the ineffaceable brand ensured, the plea could only be used once by any single individual.

Toward the end of the eighteenth century such punishments began to seem needlessly painful, although there was still time to brand a few Loyalists before penal reform was accomplished. Great Britain officially abolished branding in 1779, and most American states enacted similar legislation when their governments were established just after the Revolution. Pennsylvania, for instance, ended the practice of branding for adultery and fornication in 1791. By the beginning of the nineteenth century, branding—along with most of the other forms of corporal punishment—had been replaced by imprisonment in most legal codes throughout the country. The practice was not expunged entirely from the American scene, however. Branding had always been used by slave owners as a means of control. William Byrd, in his revealing account of life on a Virginia plantation in the eighteenth century, describes one occasion when his wife, in a fit of anger, chose to punish a housemaid by burning her with a hot iron. Unofficially, the pernicious habit continued well into the nineteenth century.[77]

Burning Alive

Burning at the stake enjoys a curious reputation in American popular culture. In literature and plays, it crops up most often as the favored means of executing witches, saints, and the innocent white victims of Indian rampages. It is used for dramatic effect and, inexplicably, is often portrayed as virtually painless. The climax of Mark Twain's *Personal Recollections of Joan of Arc*, for instance, depicts a romantic stake scene in which the young Joan lifts her face heavenward and prays fervently amid the smoke and flames until, at length, a "swift tide of flame burst upward, and none saw that face any more nor that form, and the voice was still."[78]

The historical truth contrasts with this potent mythology on a number of key points. Although Indian stake burnings occurred sporadically throughout the colonial period,[79] the burning of witches never caught on in America as it did in Europe. Even at Salem, not one of the twenty individuals executed in 1692 for witchcraft was burned (nineteen were hung and one was pressed to death). Neither were heretics routinely set ablaze, although there is little question that some Puritan ministers seriously considered that option for the Quakers. One zealot went so far as to volunteer to "carry fire in one hand and faggots in the other, to burn all the Quakers in the world."[80] As for the implication that victims of burnings accepted their fate with nobility, it is sheer fabrication. A more realistic interpretation comes from Pennsylvania's court records for 1731 concerning a woman sentenced to be burned alive for the murder of her husband. She was supposed to be hung before being burned, but somehow the fire "broke out in a stream directly on the rope round her neck, and burnt off instantly, so that she fell alive into the flames, and was seen to struggle therein."[81]

Early Americans knew from experience that being burned was a painful proposition. That is why they often stipulated that criminals sentenced to burning be strangled first. Yet it was the extreme painfulness of this ancient mode of execution that also accounts for its survival on American shores.

Children could imagine the terror and pain associated with being burned at the stake, as depicted in this early nineteenth-century woodcut. From John G. Rogers, *Specimen of Printing Types from the Boston Stereotype Foundry* (1832).

To punish perpetrators of particularly odious crimes in a dramatically public way, there was no more effective technique. Thus, the man who reportedly killed and ate his wife in Jamestown during the winter of 1609–1610 was punished by burning. More often, however, it was wives who were burned to death for the murder of their husbands. This custom was well established in British law and reflects the belief that it was unseemly to hang women in public. The notion survived the transatlantic migration with the result that similar sentences were occasionally imposed by colonial courts.

It would be misleading, however, to imply that burning at the stake was widespread in America. It was not, and by the end of the eighteenth century it had become almost unthinkable as various state constitutions and ultimately the federal Constitution expressly prohibited "cruel and unusual punishments." In a sense, the burning of effigies and other symbolic tokens can be interpreted as society's effort to find a socially acceptable alternative for burning real people. Long featured in European folk festivals, the torching of effigies figured prominently in political protests of the Revolutionary era and surfaced again during the nineteenth century as, for instance, when the volunteer firemen of Cincinnati burned in effigy the hated councilman who had helped introduce steam pumpers to the city.

This woodcut depicts Lucretia P. Cannon, murderess and head of a kidnapping gang in Delaware in the 1820s and 1830s, throwing a black child into the fire. From *Narrative and Confessions of Lucretia P. Cannon* (1841). Courtesy of the Historical Society of Delaware.

Cross burning, one of the most recognizable symbols of hatred in our own day, was primarily a twentieth-century development.[82]

Nevertheless, there was a dark corner of American society in which actual burnings had particular tenacity, namely, the world of the slave. Perhaps it was because slaves had no property and most punishments were of necessity corporal in nature. Perhaps the burning of flesh was deemed appropriate retribution for the countless burnings of slave owners' houses. Or perhaps it was sheer cruelty. In any case, burning alive was stipulated all too often as the punishment for slaves who were convicted of felonies, particularly of murdering their masters. Slaves caught participating in revolts were also vulnerable to this form of public execution, as the New York uprising of 1712 demonstrates. A group of about two dozen slaves set fire to a building and then killed or wounded a number of whites as they hurried to extinguish the flames. For this outrage, furious city officials executed eighteen slaves, four of whom were burned to death—one over a "slow fire that he may continue in torment for eight or ten hours." The governor of New York considered this to be a "most exemplary punishment."[83] Not surprisingly, when slaves revolted in New York again in 1741, retribution was similar: thirteen blacks were burned, along with a number of other rebels who were hung.

Slaves who committed felonies in the South were generally tried in special courts made up of a justice of the peace and two slaveholders. In such a setting it was relatively easy to perpetuate burning as a form of punishment well into the nineteenth century. Thus, when authorities in Wayne County, North Carolina, discovered a plot to poison and enslave whites in 1805, one of the perpetrators was sentenced to execution by burning alive. A year later, when two Negroes killed an overseer in Georgia, the first was hanged but the second was tied to a tree and set on fire.[84]

Outside the law, the pattern continued even longer. Angry mobs, swept along by racial hatred, seized and occasionally burned blacks. During New York City's draft riots of 1863, rioters not only torched buildings containing blacks but also burned the mutilated corpses of black men and women. As late as 1893 a mob in Paris, Texas, lynched and burned a mentally retarded black man who had killed a young girl. A five-day killing rampage in Georgia in 1918 resulted in a pregnant black woman's being roasted alive, but not before her baby was cut out and trampled to death.

Mankind's inhumanity toward fellow human beings has taken many forms over the millennia. That fire has been used in unsavory ways is probably less surprising than our collective amnesia of such a heritage.

Arson

When the British traveler Frederick Marryat visited New York City in the 1830s he, like almost every other tourist of the day, remarked upon the disturbing frequency of fires. According to Marryat, there were about three or four fires every day—which meant that one out of twenty houses burned to the ground each year. Intrigued by the subject, Marryat tried to ascertain the reasons for this appalling situation, and in the process he discovered something even more unsettling than the city's poor fire record: a huge proportion of the conflagrations were set on purpose. With information gleaned from the local citizenry, Marryat devised a four-point tabulation of the principal causes of fires in New York:

These fires are occasioned—

1st. By the notorious carelessness of black servants, and the custom of smoking cigars all day long.

2nd. By the knavery of men without capital, who insure to double and treble the value of their stock, and realize an honest penny by setting fire to their stores. (This is one reason why you can seldom recover from a fire-office without litigation.)

3rd. From the hasty and unsubstantial way in which houses are built up, the rafters and beams often communicating with the flues of the chimneys.

4th. Conflagrations of houses *not* insured, effected by agents employed by the *fire insurance companies*, as a punishment to some and a warning to others, who have neglected to take out policies.[85]

Marryat later discovered that New Yorkers also set fires to cover up crimes. Just before his arrival, there had been a notorious case of this sort involving a young man who, having tired of the demands of his mistress, murdered the woman with an axe. Fearing discovery, he set fire to the bed where the corpse lay and made his escape. "A merciful Providence" intervened, however. The flames were extinguished in time to reveal the crime, and authorities ultimately apprehended the criminal.[86]

New York, notwithstanding its unique size and character, held no monopoly on arson in the antebellum years. During this same era, rowdies were observed tossing fireballs at circus tents in the Midwest, and a crazed incendiary threw a pail of burning coal oil into a store in Baltimore. Arson was suspected in the huge Pittsburgh fire of 1815, as well as in conflagrations that consumed major portions of Lebanon, Tennessee, and Columbia, California, in 1854. In its early years, San Francisco was particularly vulnerable to incendiarism. One of the first duties performed by the newly established fire department in 1847 was to hang seven men for setting a building on fire. Such measures were apparently insufficient to solve the problem. Three years later the city lost 150 buildings valued at $500,000 to fires set by arsonists. According to one estimate, fully one-quarter of the city's fires between 1856 and 1860 were set deliberately,[87] a fact that would not have surprised David D. Dana, a contemporary observer. In his view, San Francisco's fire problems stemmed from its overabundance of "regularly-formed groups of desperadoes, who would as readily burn the city, murder, rob and steal, as eat."[88]

Other cities had their own groups of desperadoes. Between 1830 and 1850 there were at least thirty-five large-scale riots in the major cities of the East Coast, and in many of these episodes fire was used as a weapon along with stones, bricks, guns, and fists. Occasionally, the combatants adopted the traditional view of fire as a purifying agent and targeted their blazes carefully in order to rid society of some perceived evil. Thus, the citizens of Vicksburg, waging an emotional battle against gamblers in their community in 1835, lynched five of the wrongdoers and then dramatically burned their faro tables in a public bonfire. About twenty years later, elite volunteer companies in Chicago launched a similar campaign against prostitution and burned the entire red-light district to the ground.

More commonly, antebellum riots were disorganized confrontations between various ethnic, religious, or political groups, all of whom appreciated the power of fire to cause pain and suffering. Philadelphia's Nativist Riots of 1844 pitted native-born Protestant Americans against Irish im-

An angry mob set fire to Pennsylvania Hall on the night of May 17, 1838. Firemen made no attempt to extinguish the blaze but did protect adjacent buildings from the spread of the flames. Courtesy of The Library Company of Philadelphia.

migrants and resulted in thirty deaths and the destruction by fire of more than thirty houses and two Catholic churches. Recalling the destruction years later, one contemporary lamented that even as the beautiful St. Augustine's Church smoldered, its elegant cupola a blazing mass of wreckage in the street, the mob kept the fires of hatred alive by feeding sidewalk bonfires with books taken from the church library and furnishings seized from neighboring houses. Similar excesses in the name of religion led to the infamous torching of the Ursuline Convent in Charlestown, Massachusetts, in 1834 and the burning of more than twelve Irish houses and at least five Irishmen in Louisville in 1855. Fire also figured in the battle plans of mobs during Philadelphia's Election Riot of 1834, Baltimore's Anti-Bank Riot in 1835, and New York's Astor Place Riot in 1849.[89]

In the decades before the Civil War the inflammatory prognostications of the abolitionists consistently drew fire in a literal sense. Countless white leaders and editors were systematically burned out by opposing forces. Blacks also suffered from these hostilities, as indicated by the fate of the Reverend Peter Williams, rector of St. Phillip's African Episcopal Church in New York City. During the brutal anti-abolition riots of 1834, his church was almost totally demolished, his organ destroyed, and his furniture burned in the street.[90]

Fire, in fact, was the classic weapon in the longstanding undeclared war between blacks and whites in America.[91] Individually, servants and slaves punished their masters for various wrongs by surreptitiously setting fire to

101

barns or crops. Fire was even more effective when used by groups of slaves to distract their owners during attempted escapes or, more ominously, to stage full-scale insurrections. This application of the flame had undergone ample testing during the eighteenth century. Fire figured prominently in New York City's slave uprising of 1712, as well as in the more destructive Negro Plot of 1741. The technique surfaced again in 1811 when a large band of Louisiana slaves burned four or five plantations during a revolt near New Orleans. Likewise, during a slave rebellion in Texas in 1860 nine cities suffered major fires, including a wildly uncontrollable blaze that destroyed most of the business district of Dallas.

Terrified whites, well aware of their vulnerability, fought fire with fire. Like the slaves they feared, they frequently used arson on a small scale to "solve" local problems. From time to time, however, their confrontations were large enough to make lurid headlines. In Cincinnati in 1829, in an obvious effort to discourage black settlement in the area, newcomers were given sixty days to post individual bonds of five hundred dollars; but before the time was up, angry white mobs attacked. They killed and burned randomly in the black ghetto and ultimately forced a thousand people to move to Canada. In Philadelphia, anti-Negro riots involving arson occurred in 1834, 1838, 1840, 1842, and 1848.[92] As was all too common elsewhere, Philadelphia's mobs tried to prevent firemen from extinguishing the blazes, even if that meant cutting hose lines or attacking the firefighters directly.

Throughout the turbulent first half of the nineteenth century arson was, as it always had been, against the law. Although British common law viewed arson as simply the malicious burning of another person's house, a wide range of local American statutes ultimately expanded the definition to reflect more accurately the many horrors of the crime. By the end of the eighteenth century, arson in most states had come to include the willful burning of any house, including one's own, as well as the burning of public buildings, factories, uninhabited dwellings, and the contents of buildings. Subsequent legislation incorporated numerous additional provisions, including stern measures to deal with deliberately set fires that caused loss of life.

Punishments, though variable, reflected a similar concern for the seriousness of this crime. In most colonies during the early days of settlement, arson was a capital offense. This extreme penalty gave way only gradually to hard labor and whipping (introduced into the unusually humane Pennsylvania legal code at the end of the seventeenth century) and imprisonment and fines (the standard responses to the crime by the nineteenth century). Reflecting the widespread fear of insurrection in the South, legislation dealing with arson committed by slaves remained extremely harsh until the Civil War. Legislators in Georgia designated the burning of grain by slaves a capital crime early in the nineteenth century. In Virginia, in a similar campaign to tighten the slave codes after 1800, penalties for burning stables, barns, stacks, ricks, and houses were made stricter than ever by a special act passed in February 1808. The records show that between 1789 and 1810 at least seven slaves were executed in that state for arson, but the number was probably much higher. Vestiges of this severity endured for decades. Setting fire to a hayrick was still a capital offense in Maryland in the 1890s, and as late as 1960 at least two American jurisdictions retained the death penalty for arson.[93]

Writing the laws was the easy part. Catching and convicting an arsonist was something else again. Part of the difficulty was that evidence had a nasty habit of burning up. Even when telltale marks did survive, local fire departments generally lacked the time and expertise to investigate suspicious blazes. Equally problematic was the fact that arsonists could operate under cover of darkness (an advantage the legal system tried to counteract by imposing harsher penalties for arson committed at night). Additionally, in a world plagued by accidental conflagrations, arsonists had almost unlimited options for simulating household accidents. One trick was simply to place a lighted candle or lamp near a window where the breeze would blow the curtain against it. Another approach was to use animals to expedite ignition. For instance, it was not uncommon for hired arsonists to tie a lamp to a cat's tail; as soon as the cat jumped, the lamp would automatically overturn and start a blaze. In at least one documented case, a firebug is known to have tied a piece of meat to a gas light so that it hung down over a lighted lamp. The villain then introduced a hungry dog into the room and departed, fully confidant that the dog would jump for the meat, upset the lamp, and start a fire.[94]

Faced with such malicious cleverness on the part of arsonists, authorities searched continually for measures that would cut down on the crime. In Philadelphia Mayor Richard Vaux, a former fireman, established a special investigative unit in 1856. Over the next few decades, officials in other jurisdictions called for similar legislation. "In all municipalities there should be a special police charged with the duties of discovering the origin of fires," declared one concerned Ohio official in 1881.[95]

Monetary rewards aided in the apprehension of arsonists. This pattern was well established on the municipal level as early as 1723 when the lieutenant governor of Massachusetts offered a reward of fifty pounds for the conviction of Boston firebugs.[96] By the third quarter of the nineteenth century insurance companies had adopted a similar approach. Between 1873 and 1881 a nationwide group of insurers, operating under the aegis of the Committee on Incendiarism and Arson of the National Board of Fire

Underwriters, offered a total of $429,000 in reward money for the arrest and conviction of arsonists.[97] Leaving no stone unturned in the fight against incendiarism, one Massachusetts town even tried the unusual method of mesmerism to aid in the capture of a local firebug. "Knaves may well dread the approach of mesmerism henceforth," warned *Scientific American* in 1846.[98]

With the exception of the hypnosis technique, these measures had some success. Mayor Vaux claimed that his bureau in Philadelphia cut the incidence of arson in that city by half. The insurance companies also saw results from their incentive program. Between 1876 and 1881 the Fire Underwriters commission paid out $19,729 in sixty-five separate awards. In the process, 108 arson convictions were rendered, 9 of which carried life sentences. Pessimists, however, did not have to search far to find persuasive evidence that arson was as serious a problem during the second half of the nineteenth century as it had been during the first. If the Fire Underwriters awarded only $19,729 of $429,000 available during the period 1873–1881, then it is clear that many cases of arson were never adjudicated. Just how many is hard to say, but insurance companies attempted periodically to come up with useful statistics. According to one of their tabulations, 15 percent of the fires reported outside major U.S. cities in 1880 were of incendiary origin. Economic losses from this source amounted to 19 percent of the total fire losses for that year.[99]

Documentary evidence corroborates the persistence of the problem. The New York City Draft Riot of 1863 resulted in the deaths of at least three hundred people and the torching of scores of buildings, including a mission, an orphanage, an armory, and three police stations.[100] Arson was also a recurring factor in such labor confrontations as the railroad strike of 1877 and the Pullman strike of 1894, and it figured sporadically in the so-called county-seat wars in which overzealous town boosters of the Midwest and West fought with their neighbors over the location of the county courthouse.[101] Race relations continued to deteriorate in the lurid glow of the arsonist's torch. Throughout the Reconstruction period in the South, houses and churches belonging to blacks were burned to the ground by whites seeking to reassert their dominance. In some localities, schools erected by the Freedmen's Bureau were destroyed by fire as soon as they were built.

These are, quite literally, textbook examples of mob violence and arson. It was, rather, the average, "everyday" incendiary who deserves the real credit for spreading the crime throughout the country and transforming fire from a benevolent source of energy into a dangerous tool of destruction. The motives of these characters varied enormously. A discharged coachman set fire to Abraham Lincoln's stables in 1864 to exact revenge. A group of Chicagoans, disappointed when the great fire of 1871 was extinguished, started new blazes in order to perpetuate the opportunities for looting. Daniel Smithson torched a church in Greensburg, Pennsylvania, in 1880 in order to cover up his theft of the church's carpet and clock. George R. Miller of Meriden, Connecticut, burned his own store in the same year for the insurance.

Occasionally, the motivations of these arsonists were so unconventional that their cases merited extensive public comment in contemporary publications. From such a source we learn that a Pennsylvania man had a propensity for burning barns and houses simply because "he thinks no one is entitled to have more property than himself." Equally bizarre is the story of an impoverished Wisconsin barn burner who, as the local paper put it, "committed the crime for the sole purpose of being sent to prison where he would at last receive enough to eat." Not infrequently, arsonists had no reason for their actions other than the sport of it. The *Rochester Herald* reported with satisfaction the arrest of two young men who "had already burned down a house on Orchard Street and a stave factory when they were arrested, and they had planned to fire three more Wednesday night if the detectives had not spoiled their game." The reporter noted that one of the boys "was inspired to his criminal work" by reading "abominable" dime-store novels and concluded that society "ought to take some steps to suppress the demoralizing stuff."[102]

Such action would not have stopped Lydia Berger, who set her father's barn on fire in 1898 as revenge for a whipping. Nor would it have influenced Anna Myinek, who set fire to her employer's house and barn two years later because "she was lonesome and homesick, and wanted excitement."[103] Mysterious fires would still have broken out in Laurel, Delaware, in 1885, in Frog Point, Michigan, in 1912, and in any number of other localities throughout the country because, as everyone knew, novels did not cause arson—human nature did. Like arsonists of today, incendiaries of the nineteenth century displayed a number of different behavioral types; but in most cases—whether the crime was committed for revenge, profit, cover-up, or excitement—the basic motivating force was power. All men were not created equal, but with the simple flick of a match they could be rendered more so. During the seventeenth century, Americans referred to arson as the "the workman's trick," a clear reference to the ease with which a laborer could retaliate against an employer for a perceived wrong. Centuries later, William Faulkner evoked the same idea in "Barn Burning," a brooding story in which the protagonist confronts an enemy with the menacing message, "Wood and hay kin burn." Just as fire powered steamboats and railroad engines, it also empowered people, who consistently saw in the flame a sure-fire way to right a wrong, to gain wealth, and to inject a little excitement into ordinary lives.

To view arson as a means of strengthening the weak almost serves to legitimize the crime, but it should never be forgotten that the victims were everyday people too.[104] In 1880 alone, Americans suffered reported losses of $4,815,000 from arson.[105] Almost every type of home and workplace was at risk. So was people's peace of mind. Although there is no way to assess the additional psychological damage caused by threats of arson, diaries occasionally refer to this dark subject, particularly during times of war, when soldiers torched private houses or, as Mary Chesnut observed during the Civil War, when mutinous slaves threatened to burn their masters' property. Luzena Stanley Wilson, however, had a brush with arson during peacetime, in Nevada City in 1849, and her account reveals some sense of the terror generated by an angry mob with fire on their minds:

> One night I was sitting quietly by the kitchen fire, alone. My husband was away at Marysville, attending court. Suddenly I heard low knocks on the boards all round the house. Then I heard from threatening voices the cry, "Burn the house." I looked out of the window and saw a crowd of men at the back of the house. I picked up the candle and went into the dining room. At every window I caught sight of faces pressed against the glass. I hurried to the front, where the knocking was loudest and the voices were most uproarious. Terrified almost to death, I opened the door, just enough to see the host of angry, excited faces and hear the cries, "Search for him" and "No, no, burn him out." I attempted to shut the door, but could not.[106]

Wilson soon determined that the mob was searching for a murderer who reportedly lived in her boardinghouse. With great presence of mind, she allowed the men to search the premises and thus saved her home.

Other victims were not so lucky.

WHEN THE FLAME GOES OUT

The world of our forefathers could be a bone-chilling place. "'Tis dreadful cold," Cotton Mather wrote in his diary one day during the searingly cold winter of 1719–1720. "My Ink-glass in my Standish is froze & splitt, in my very stove. My Ink in my very pen suffers a congelation: but my witt much more."[107]

With the passage of time came improvements in heating devices but hardly elimination of the problem. Ellen McGowan Biddle, recalling her experiences as a soldier's wife at Fort Whipple, Arizona Territory, in 1876, describes living in rustic military quarters that had sizable cracks in the walls. "As there was often in winter a difference of fifty degrees in temperature between the day and night-time," she explained, "we had to keep great fires going continually. We had no stoves or furnaces, only the large open hearth fire, and it is needless to say it was hard at times to keep warm."[108] It was so hard to keep warm in Iowa during the great blizzard of 1863 that children were reportedly kept in bed for days on end as protection against the frigid air in their houses.[109]

American history offers numberless chilling tales, each of which serves to demonstrate not only that physical fortitude was a virtue but also that on certain occasions the worst type of fire was no fire at all. It is not by accident that Dante's Ninth Circle of Hell is a lake of ice rather than a flaming pit. Nor is it surprising that Samuel Thomson, the founder of a radical approach to medicine during the first half of the nineteenth century, proclaimed that all diseases were caused by coldness and could be cured by the remedial application of heat. To be without fire in the cold was to suffer—sometimes unbearably.

Professional writers were particularly adept at describing such deprivations. The winter hardships of the Pilgrims at Plymouth and the Continental Army at Valley Forge were favorite themes of nineteenth-century authors, but their own century offered powerful examples of its own. In *White Jacket*, for instance, Herman Melville captures the shivering mood on board a naval frigate of the 1840s:

> Oh! the chills, colds and agues that are caught. No snug stove, grate, or fire-place to go to; no, your only way to keep warm is to keep in a blazing passion, and anathematize the custom that every morning makes a wash-house of a man-of-war.[110]

Louisa May Alcott includes a similar scene of cold and suffering in *Little Women*, and Jack London's "To Build a Fire" raises the subject to the level of metaphor.

The impact of such literary evocations was heightened by the fact that cold fires and extinguished flames were recognized symbols of death. Nevertheless, these passages were firmly rooted in reality and were therefore uncomfortably similar to real-life situations. As early as 1637 the indigent population of Boston suffered terribly from scarcity of firewood. Not infrequently, they simply made do without a fire on the hearth.[111] Although local histories tend to gloss over the subject, this problem survived the colonial period and reappeared periodically throughout the nineteenth century. Regional fuel shortages, such as the crisis that afflicted Philadelphia during the War of 1812, exacerbated the problem; but even in prosperous times families without money tended to be families without fire. Foraging for sticks of firewood, searching for stray pieces of coal near the railroad, and begging and stealing were the unenviable lot of the most desperate. Although such measures often sufficed to carry a household through a crisis, many of the nation's poor also depended on the charity of their neighbors. The following verses from a mid-nineteenth-century poem are typical of many efforts to remind the general public of the plight of the poor during the winter and to shame them into lending assistance. Fire colors the rhetoric even as it represents the problem:

"O, ho! O, ho!" quoth old Jack Frost
As he looked at the farmer's fireside,
   And saw the huge log
   On the bright fire-dog,
And a flagon and tankard beside;
And heard the gay jest, and the loud merry laugh,
   As they trill'd forth their Christmas rhymes,
How happy they grew as the liquor they'd quaff—
   Jack gave a huzza for old times!

"O, ho! Oh, ho!" sighed old Jack Frost,
As he look'd in the poor man's hut,
   Dark, dirty and drear,
   And no fire to cheer,
Without window or door that would shut!
And a half clad mother her children cuddled,
   To give warmth to the nestling brood;
And the little ones cried, as together they huddled,
   "Oh, mother, pray give us some food!"[112]

If all went well, the plight of this family would be temporary, their suffering eased by good neighbors, good fortune, or a change in the weather. There was another possibility, however. "Here lie Paul, Rachel, Amos, John and tiny Richard, put to an early death by the misability of sister Elizabeth to light a fire in the hearth," reads the epitaph at the grave of a family that froze to death during the last century.[113] Less public, but equally revealing, are the many journalistic accounts of similar tragedies.

MEDICAL CURES FOR BURNS

In slight burns which do not break the skin, it is customary to hold the part near the fire for a competent time, to rub it with salt or lay a compress upon it dipped in spirits of wine, brandy, or cold vinegar. But when the burn has penetrated so deep as to blister or break the skin, it must be dressed with some of the liniment for burns, mentioned in the Appendix, or with the emolient and gently drying ointment [called] *Turner's Cerate* [a mixture of olive oil, wax, and calamine]. . . . When this ointment cannot be bad [sic] an egg may be beat up with about an equal quantity of the sweetest salad oil.

William Buchan, Domestic Medicine, 1797

If a person who is burned will *patiently* hold the injured part in water, it will prevent the formation of a blister. If the water be too cold, it may be slightly warmed, and produce the same effect. People in general are not willing to try it for a sufficiently long time. Chalk and hog's lard simmered together are said to make a good ointment for a burn.

Lydia Maria Child, The American Frugal Housewife, 1833

In slight or moderate burns, and sunburn, the pain is best relieved by wrapping the parts in cloths wet with a solution of ordinary baking soda. As much soda should be dissolved in the water as the water will take up, which is called a saturated solution. In severe burns the clothing should be removed with great care so as to avoid tearing the skin where it adheres; no violence should be used; . . . disinfect the parts with warm boric acid solution; then dust with boric acid solution . . . ; in the dressing, the great object is to exclude the air and for this purpose a free use of carbolized vaseline is excellent; smear it thickly on the parts and then dress with cotton which has been sterilized by prolonged heating or with sterilized gauze. . . . If healing is slow a weak solution of sulfate of copper, ten grains to the dram of boiled water, may be used to stimulate granulations.

The Cottage Physician, 1900

Spit on a burn to cure it.

Rub grease on burns to cure them.

If you burn your finger, stick it in your ear and the soreness will go away.

Suck a burn to draw out the fire.

Boil about fifty strong red pepper pods. Strain the solution through cheese cloth.

Reduce this to a small quantity by heating it, then pour it into a pint of melted buffalo tallow. Lard is nearly as good as buffalo tallow. The pain will leave five minutes after the application of this remedy.

Welsch, comp., A Treasury of Nebraska Pioneer Folklore

Fighting Back

In 1851 a small group of American businessmen decided that they had had enough of the "fire-fiend" and its "immense destruction." They decided to fight back. Living as they did in the golden age of invention and optimism, these entrepreneurs placed their faith in an intriguing new device called the Phillips Fire Annihilator. Unlike conventional firefighting equipment, this mechanism used gas rather than water to destroy unwanted flames. The gas was produced in a sturdy iron cylinder by combining sulfur, saltpetre, and a third "secret" component with controlled amounts of sulfuric acid. The resulting chemical reaction produced a small explosion that forced the newly created gas out of the container and onto the fire.

Patented in England, the invention quickly made its way to the United States, where an investors' consortium extolled its virtues to the public. As their advertising brochure pointed out, the Phillips Fire Annihilator was safe, effective, and—most attractive of all—portable: "You can have it in your chamber, your parlor, your vault, your treasure-room, your library, or your kitchen. And no matter where you have it, it acts like a protecting spirit." Newspaper testimonials echoed this praise. One even predicted that "the time is not far distant when every house will have a 'Fire Annihilator,' and such a thing as a serious conflagration will be unknown in our country."[1]

Such claims sounded too good to be true, but the promoters were prepared for skepticism. With the confidence of salesmen and the bravado of showmen (P. T. Barnum was one of their number), they staged public demonstrations in which buildings were deliberately set on fire and the annihilator put into action. Unfortunately, however, the results were not always satisfying, and even the carefully controlled experiments occasionally flared out of control. On December 18, 1851, for instance, although the annihilator successfully passed its test, some spectators—"rowdies who were opposed to the invention," Barnum claimed—torched the demonstration building shortly after the exhibition. This time it burned to the ground. On another occasion, when a wooden building at the U.S. Navy Yard was ignited, the annihilator had no effect on the flames and the building was quickly destroyed. These and other documented failures

The Phillips patented fire extinguisher was one among dozens of chemical extinguishers introduced in the second half of the nineteenth century.

proved what the editors at *Scientific American* had been saying all along: that although the Phillips Annihilator "might do good and be useful in many cases," the promoters of the apparatus had shamelessly "*claimed too much.*"[2] They had pursued a single cure for the national fire problem when, in fact, there was none.

This disappointing reality did not prevent enterprising entrepreneurs from hoping. For decades they courted the public with gadgets that ranged from the "very perfect" Eclipse fire engine to the peerless "Shur-Stop" fire grenade. Sometimes they even enjoyed limited success. Nevertheless, the cure-all remained elusive, and the country continued to rely, as it always had, on a multidimensional fire campaign that included fire prevention, firefighting, and fire insurance. Throughout the nineteenth century, citizen action on all these fronts was primarily reactive: a fire spurred limited action of some sort, and successive blazes spurred refinements of the response. In addition, such action was preeminently local.

PREVENTION

Probably nowhere has the dictum "There ought to be a law" been more assiduously applied than in the field of fire prevention. Plymouth Colony signaled the trend as early as 1627, when, in the wake of several serious house fires (one of which almost killed Governor William Bradford), officials enacted legislation prohibiting thatched roofing. Four years later, Boston's selectmen banned thatching as well as wooden chimneys, both of which were common at the time. When a fire destroyed more than 150 wooden buildings in Boston in 1679, the General Court tightened the building code by ordering that all new houses be constructed of "stone or bricke, & covered with slate or tyle."[3]

Throughout the colonial period, other fire hazards were quickly identified and just as quickly dealt with by local authorities. The most common ordinances regulated chimney cleaning, underbrush burning, and the storing of gunpowder, but many jurisdictions also prohibited smoking tobacco "out of dores," lighting bonfires in populated areas, and carrying lighted brands in urban areas without adequate protection. Philadelphians were not permitted to keep hay in their kitchens, and Bostonians were not allowed to light fires on vessels docked at the wharf. The curfew, Medieval Europe's traditional method of reducing nocturnal fires, was also introduced in the colonies at an early date. As the French derivation of the word implies (*couvre* and *feu*), householders were compelled to cover their fires at night. In Boston's version, adopted in 1649, all fires had to be covered or extinguished between 9 P.M. and 4:30 A.M.

These colonial ordinances provided the groundwork and inspiration for many of the fire statutes of the nineteenth century. Building codes, in particular, were strengthened throughout the century, most notably in the larger cities, where the typical ordinances gradually evolved beyond simple requirements for fire-resistant construction of brick and masonry to more precise specifications for roofing material, building height, and wall thickness. The great fires in Chicago, Boston, and Baltimore prompted additional tightening of the codes in those localities, although legislation in all three places fell far short of the possibilities. Boston's response was perhaps as advanced as any: stiff requirements for brick barriers between floors; "more fully" fireproofed boiler rooms, stairs, and elevators; and elimination of wooden roofing and trim.[4] Naturally, the nation's fire underwriters watched these and other local developments carefully, and in 1896 they tried to hasten progress by joining with other interested parties to form the National Fire Protection Association. This organization, in conjunction with citizen groups and informative publications, helped to oversee the campaign that resulted in the standardized building codes of modern times.

Chimney fires, which continued to threaten densely settled areas throughout the 1800s, also prompted endless rules and regulations. Ordinances focused on the composition and height of the stacks, as well as on the cleanliness of the interior walls. Methods of inspection and punishment varied, but the approach adopted at mid-century by Mobile, Alabama, was typical:

> Be it ordained by the Mayor, Aldermen and Common Council of the city of Mobile, That it shall be the duty of each owner or occupant of every house within this city to sweep, or cause to be swept, at least once a month, any chimney where he, she or they habitually keep a fire; and if any chimney shall take fire through neglect of being properly swept and cleaned, the occupant of the house, room or apartment to which such chimney appertains, shall forfeit and pay the sum of five dollars.[5]

Many communities exerted additional control over the chimney-cleaning process by awarding monopolies to reputable sweepers and by regulating rates and performance standards.

Thus, fire prevention in the nineteenth century was to a great extent an elaboration of earlier practices. Yet the nineteenth century was a time of extensive technological innovation, and many of the era's most successful inventions spawned fire hazards that had no precedent in the colonial period. Local authorities tackled these new dangers as enthusiastically as they had the old, producing in the process a crazy-quilt of laws that reflected the concerns of individual communities. Officials in Farmington, Connecticut, for example, forbade the use of portable stoves and required householders to store their matches in iron boxes. Philadelphia's councilmen were more concerned about wood-burning locomotives, which, as contemporary wags put it, often seemed to burn more wood *outside* their fireboxes than within. After some debate, these spark spewers were summarily banned from the city limits. Pittsburgh's officials regulated stovepipes and bonfires, while their counterparts in Greenfield, Indiana, prohibited the "lighting or burning mischievously of any shavings, wood or other rubbish." Although unlighted streets were an obvious civic problem for Cincinnati, the carelessness with which private citizens erected their own street lamps quickly posed more of a threat than darkness. In 1815, therefore, the city council forbade unauthorized street lights and imposed an unusually steep fine of fifty dollars on transgressors.[6]

Theater fires presented a perennial challenge to lawmakers. The stringent regulations passed in the wake of the Brooklyn Theatre fire (1876) and tightened after the Iroquois Theater fire (1903) represent well-intentioned attempts to combine fire prevention measures—including requirements for the use of flameproof materials and periodic inspections—with strict safety codes—more and better marked exits and fire escapes.

But, admirable as such efforts were, the limitations of the regulatory approach were clear from the outset. It was not simply that ordinances were hard to enforce, although this was certainly part of the difficulty. More problematic was the virtual impossibility of predicting and regulating every single fire hazard in every single locality. This shortcoming might be called the "incendiary pig syndrome," in honor of a porcine mischief maker from Maryland who, as the story went, "caught hold of an apron, which it dragged to the fire, and then under the bed, whereby the flames caught and destroyed [a] house" in 1846.[7] Few legislators would have suspected pigs of arson, and almost no one would have recommended making pigs illegal.

Chicago provides a good example of this syndrome in action. Although the city had passed a number of ordinances banning wooden building materials before the 1871 inferno, no one had thought to tear up the miles of wooden sidewalks or to require demolition of existing wooden buildings or to forbid the construction of wooden sheds. Predictably, these structures were ready fuel for the great fire. Afterwards, a chastened populace tried to make up for such laxity. Indeed, many concerned citizens jettisoned partisan politics in order to support the newly formed "Fireproof Party" and its aggressive drive for improved building codes. Even these efforts fell short of the mark, however. Not only had considerable unauthorized construction already taken place by the time the postfire elections were over, but, as had happened in Boston two hundred years earlier, rigid requirements for expensive fire-resistant materials inevitably pitted the rich against the poor and, in so doing, virtually guaranteed failure.

The danger of fire could not be simply legislated away. Fire prevention tactics of the nineteenth century thus moved beyond legal constraints to supplementary avenues of support. While Americans in search of a fire-safe environment had many allies, two were especially revered: an informed public and fireproof materials.

Fire Prevention Day was inaugurated in 1911 when a number of states, at the instigation of the National Board of Fire Underwriters, designated the anniversary of the Chicago fire as a day to promote fire safety. Within a decade, a presidential proclamation had elevated the concept to a national event, and the nation's first citywide fire prevention bureau had been formed in Portland, Oregon.

The absence of such high-profile programs in earlier decades does not mean that nineteenth-century urbanites failed to take fire seriously, however. As might be expected, fire prevention measures were most conscientiously implemented in some of America's utopian communities, where the welfare of the group was a stated goal. The Harmonist commune near Pittsburgh, for instance, had an extraordinarily well-organized program during the 1870s and 1880s: "Everything in the community was kept clean and safe; absolutely no rubbish was to be seen; everything inflammable was cautiously guarded; smoking was forbidden; matches were kept in iron wall brackets with two compartments . . . and both old and young passed through a daily discipline of caution."[8] Elsewhere, safety tips were urged upon the public in a less coherent, but equally earnest, fashion by any number of writers and editors. "Most disastrous results," claimed one, "have frequently arisen from neglect of the most ordinary precautions." Some such precautions were then dutifully spelled out:

Householders cannot be too careful that matches be cautiously used. All fires should be safe and all lights extinguished at night. No combustible substances [should be] permitted so near the stoves or grates as to be in danger.[9]

Other common suggestions included keeping the stove away from the wall, cleaning the chimney regularly, placing ashes in a nonflammable receptacle, and, of course, maintaining eternal vigilance both privately and via official fire patrols.

In their widely circulated manual, *American Woman's Home* (1869), Catharine Beecher and Harriet Beecher Stowe focused on domestic fires and suggested the following: "When quitting fires at night, never leave a burning stick across the andirons, nor on its end, without quenching it." Also, "see that no fire adheres to the broom or brush [and] remove all articles from the fire." Another popular domestic adviser, Alexander Hamilton, turned his attention to the nation's workmen, particularly woodworkers, whose tendency to allow shavings and other debris to accumulate on the floor was a clear invitation to disaster. "Nine-tenths of the number [of conflagrations] originate in carelessness and inattention" of this sort, he admonished.[10]

New technologies occasioned new hazards, particularly when employed by uninformed individuals. Thus Denison Olmsted, in explaining how to use anthracite coal in 1836, warned that "no part of a stove or pipe should ever become *red hot*." John Lee Comstock, while describing new and potentially explosive lamp fuels in 1853, stated categorically, "Never trust children, or careless persons, with the use of the burning fluid." Reporting on a terrible gas explosion in Nashua, New Hampshire, the editors of *Scientific American* reminded readers of yet another safety tip. The accident was caused by "introducing a light into a cellar, where there was a strong odor of gas, caused by its escape from a leaky pipe." One can almost hear the reporter sniffing in disdain as he added, "The person who introduced the light was not a reader of the Scientific American, or he would not have acted so unwisely, after what we have said in reference to such cases."[11]

But, of course, only a small proportion of the population did read *Scientific American* or any other manual. And even those better-informed persons were easily victimized by fire. The *Scientific American* editors themselves were burned out in 1845 by a man carelessly distilling spirits in the basement of their New York office building. Clearly, as most people knew from firsthand experience, fire prevention would have a far better chance if the foolhardy habits of individuals could be counteracted by the use of foolproof materials in everyday life. It did not take long for that realization to be translated into a promising new weapon in the war on fire: fireproof construction.

The fireproof approach to hostile fire was in perfect tune with a world that believed in the perfectibility of society and viewed technology as a favored instrument toward that end. In a sense, colonial regulations regarding roofing and chimney materials were precursors of the fireproof movement, but it was not until the nineteenth century that this concept achieved the status of a crusade. As befits a grass-roots movement, fireproofing advocates actively encouraged individuals to protect their surroundings with do-it-yourself measures. From the 1840s until well into the next century, recipes for the fireproofing of everything from paper and cloth to paint, mortar, and wood appeared regularly in the intellectual and popular publications of the day. Most of these formulas relied on some combination of calcium oxide (lime) and water in order to produce a thick liquid that hardened into an incombustible crust. Fireproof paint, for instance, could be prepared at home as follows:

> Take a quantity of quicklime and slack with water in a covered vessel. Add water or skim milk, or both, and mix til like cream. Then add: 20 lbs. alum, 15 lbs. potash, [and] 1 bushel salt to every hundred lbs. of creamy liquor. For white: add plaster of Paris or white clay. Strain.[12]

For the less adventurous, commercial manufacturers offered a seemingly endless array of fireproof products. Roofing materials, paints, boiler cover-

Asbestos found applications in numerous fireproof and fire-resistant materials. Firms specializing in asbestos distributed extensive catalogs of their products.

ings, stove linings, lamp shade collars, fire escape rope, work gloves, and building felt—all were advertised widely during the second half of the nineteenth century, and all carried the "fireproof" seal of approval. The means of achieving this magical quality depended on the product and the patent, but several favored materials deserve mention. Coal tar, a by-product of illuminating gas production, was readily adapted to both fireproof roofing and road paving. Asbestos, a fibrous mineral known since antiquity for its fire-resistant qualities, turned up as the key component in paints and roof coverings, as well as in a plethora of woven artifacts, such as industrial cloth, sewing twine, and—most welcome of all—theater curtains.[13]

The lightning rod offered a slightly different approach to the fireproofing campaign. Benjamin Franklin introduced his version of the device in 1753, and ensuing decades brought constant reminders of the desirability of such a tool. *Scientific American*, for instance, decried the "terrific effects of lightning" on homes and barns in New England during the excessively hot summer of 1845.[14] Nevertheless, owing to the expense of installation, or to the resistance of insurance companies to allow better rates for buildings so protected, or, perhaps, simply to indifference, lightning rods did not achieve widespread popularity during the last century. Of the 1,847 build-

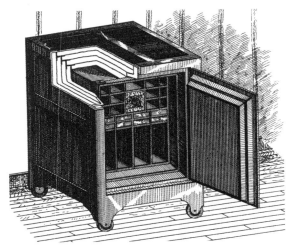

Fireproof safes featured multilayered walls and doors.
This model appeared in Benjamin Butterworth's
The Growth of Industrial Art (1892).

ings reportedly struck by lightning in 1900, for instance, only 40 had lightning rods in place. Perhaps the unpredictability of lightning strikes made such protection seem like a needless luxury.

That was hardly the case with the fireproof safe. Inventors used a variety of materials to produce fire-resistant containers that appealed enormously to homeowners and commercial establishments alike. One of the earliest models to achieve success, the Salamander Safe patented by Charles A. Gayler in 1833, boasted double walls of iron separated by air or some other nonconductor of heat. Subsequent models used asbestos, plaster of paris, fireclay, hydrated gypsum, hydraulic cement, alum, and a number of other materials as the filler. By the turn of the century, Sears was able to offer customers an improved and eminently affordable model constructed of two steel boxes separated by a filling of concrete. This "absolutely fireproof" safe came in a handy "family" size, which, like all the other models, could be shipped anywhere.[15]

Tantalizing as these many consumer products were, the real showpieces of the movement were the fireproof buildings. Taunting the power of the flame like so many children's fortresses, these edifices rose sporadically throughout the century to provide real-life testing grounds for promising new building materials. One of the earliest was Charleston's "Fireproof Building," constructed of stone and iron in the early 1820s. Later versions included New York's famous Crystal Palace, which was built of glass and iron, and San Francisco's Palace Hotel, which boasted twelve-foot thick foundation walls, a wrought-iron frame, and carefully designed stairways of brick and stone. One after another they were erected, and—much to the dismay of the public—one after another most of them succumbed to the flame. A disastrous fire in Mobile, Alabama, signaled the trend as early as 1827. "The intensity of the heat was so great and the torrent of fire so impetuous, that everything wooden or brick, fire-proof or not, disappeared before it like columns of snow," wrote one contemporary.[16] New Orleans residents watched the same fate overtake their city during a blaze in 1854. "The walls of these buildings were supposed to be fire-proof," wrote fire historian David D. Dana, "but they did not deserve the credit. Pork, lard, bacon, liquor, oil, gunny-bags, rope, and such material, when all in flames at once, are strong tests of fire-proof walls; and these speedily failed."[17]

Combustible furnishings contributed to the destruction of the Crystal Palace in 1853. Within a mere half hour of ignition, this fireproof building and the art works on display inside were turned to ashes. It took somewhat longer to consume the Iroquois Theater, Potter Palmer's hotel in Chicago, and the Palace Hotel in San Francisco, but these and countless other "fireproof" edifices were ultimately destroyed by fire.

Interestingly, what was not destroyed was faith in the notion of fireproofing. Architects and builders witnessed these failures and learned from them. Wisely altering their terminology from "fireproof" to "fire resistant," they came to understand the shortcomings of seemingly invulnerable materials like brick, stone, tile, and many metals. In response, they gradually devised a more effective approach to construction. Important innovations included the sheathing of structural iron and steel frames with ceramic tiles or cement, the use of fire-stops in ceiling construction, and the insertion of a layer of asbestos between floors. A flurry of experimentation in fire retardation techniques occurred around the turn of the century (Frank Lloyd Wright published designs for a concrete "fireproof" house in *Ladies' Home Journal* in 1907), and in time engineers arrived at the key concepts that dominate research in the field today: the notion that a building's design and construction is at least as important as its materials and the recognition that the flammability of the whole is profoundly affected by the flammability of the contents.

The general public persisted in its quest for fireproofing as well, partly out of habit and partly because there were plenty of salesman on hand to convince them that the newest product offered the opportunity of a lifetime. The great fire in Laurel, Delaware, illustrates the point. On June 24, 1899, fire broke out in the local pool hall and spread with such ferocity that sixty-two stores, twenty-eight houses, and thirty stables were burned to the ground before firefighters could control the blaze. The scene of utter devastation that greeted citizens the following morning was bleak—except for one rather comic detail. Scattered amidst the rubble and smoking ashes of the commercial district was a corps of tattered survivors, instantly recognizable as company safes. "Nearly every business house [that was] destroyed had a safe, and these receptacles are lying about, some cracked, some warped and twisted by the heat, and others intact," reported the local newspaper. Not much could be said on behalf of the warped and twisted specimens, but those that had withstood their trial by fire were advertising gold. Agents from safe manufacturing companies arrived almost immediately "to extol the merits of their firms." Two rivals even had a furious argument in the middle of the street.[18]

The results of such antics were predictable: increased sales for safe manufacturers and, more important, heightened appreciation of society's pitifully limited, but nonetheless demonstrable, ability to confine the destructive effects of fire through the use of fire-resistant materials.

FIREFIGHTING

The citizens of Laurel may have had faith in their fireproof safes, but on the night of June 24, 1899, they were thinking about another form of fire protection: firefighters. The fire alarm was sounded just after 2:00 A.M., and within half an hour wild swirls of flame had begun to sweep through the many wooden structures in the business district. Realizing that the town's hand pumper could never deal with the situation alone, Laurel's mayor sent an urgent message to neighboring communities. "We are at the mercy of the flames. Send help immediately!"

The response was more than he could have hoped for. Firemen from Salisbury, Maryland, received the distress call just after 3:00, and within twenty minutes nineteen firemen, their chief, a steamer, and one thousand feet of hose were on a special train racing toward the flaming town. They arrived at the Laurel depot eighteen minutes later and went to work immediately. Pocomoke's firemen arrived soon thereafter and quickly linked their hose and their efforts to Salisbury's. Although Wilmington was separated from Laurel by almost the entire length of the state of Delaware, that city immediately recruited two companies of volunteers and organized a special rescue train. The men and their equipment charged out of Wilmington at 4:00 A.M. and, by racing at speeds up to sixty miles an hour for two hours, reached Laurel at 6:00. The Wilmingtonians joined the battle already in progress, and by 9:15 the Laurel fire was finally brought under control.[19]

The dramatic sweep of this tale demonstrates why firefighters and their exploits have become the best-loved and most chronicled aspect of American fire history. For sheer excitement, few narratives of the past can match the rousing descriptions of red-shirted men doing battle with the flames. And for exaltation of the glories of technology, few stories can rival those that chronicle the successes of the firefighting machines of the past.[20]

Equipment

Traditionally, the history of the nation's firefighters has been presented as a tale of progress. A brief survey of the tools of the trade lends credence to this notion. As diaries and town records make clear, colonists of the seventeenth century fought fires in the old-fashioned way: they dumped water on them. Although water inflicted damage of its own, it lowered the temperature of the fuel and inhibited the flow of oxygen enough to extinguish the blaze. To aid in this process, large towns like Boston and New Amsterdam maintained a municipal supply of leather buckets; at the sound of the alarm, these were filled with water and passed via the classic "bucket brigade" toward the flames. In order to increase the amount of water, citizens were often required to contribute their own private buckets as well.

Additional implements of the early days included hooks (to enable citizens to pull down the walls of an engulfed building and stop the spread of the fire) and ladders (to enable firefighters to reach the flames on upper stories and roofs). Following a highly destructive fire in 1653, Boston officials augmented their firefighting arsenal by ordering each householder to acquire both a ladder "that shall rech to the ridg of the house" and a "pole of about 12 feet long, with a good large swob at the end of it, to rech to the rofe . . . to quench fire in case of such danger."[21] Gunpowder, used to blow up burning buildings and thus halt the spread of fire, was a more radical weapon used by seventeenth-century firefighters.

The inadequacies of such measures were clear to everyone from the outset; but it was in Boston, where unusually severe fire losses had occurred in the early years, that citizens decided to do something about the problem. In 1679, after yet another fire had savaged the town, Bostonians imported a fire engine from England. It was a rustic affair—a rectangular wooden tub surmounted by a pump and a nozzle—but it pointed the way to the future. Suddenly city dwellers had a firefighting device that promised to deliver more water with greater force and precision than conventional buckets ever could. Not even Paris could boast such an ally at this point.

Boston's fire engine was a rudimentary device—pumped by hand, filled with water by hand, and carried to fires (via long poles) by hand. Innovations came quickly, however. When the citizens of Charleston, South Carolina, imported an engine in 1713, their version came equipped with wheels as well as double-action handles that permitted a number of men to

These New York firemen are shown parading with their old model pumper on July 5, 1851. Woodcut from *Gleason's Pictorial Drawing Room Companion*.

pump together. Richard Mason, one of the early American fire engine builders, achieved an improved flow of water in 1768 by placing the pump handles (called brakes) at the ends of his engines rather than along the sides. Modifications continued on both sides of the Atlantic until, by the 1840s and 1850s, fire engines could employ the muscle power of twenty or thirty men together to produce a steady stream of water 150 to 200 feet long. John Agnew exceeded even these statistics with an engine that tapped the strength of forty-eight men, and in 1845 an innovative New Yorker patented a machine that made room for a hundred workers.[22] By that time, however, fire engine design had taken a new direction, and the public was being introduced to prototypes that called for fewer men, not more.

The revolution that made this innovation possible was the steam engine. It is generally acknowledged that the first steam fire engine was constructed in London in 1829 by George Braithwaite in consultation with John Ericsson. It took a few years for their efforts to be duplicated in the United States (New York City fire insurance companies commissioned a lumbering seven-ton version in 1841) and even longer for a successful model to reach the attention of the public (Cincinnati engineers Moses, Alexander, and Finley Latta finally produced a reliable engine for that city in 1852). But from mid-century on, despite the objections of legions of soon-to-be-displaced pumping firemen, the ascendancy of steam power

Steam fire engines greatly improved firefighters' ability to pump a strong stream of water at fires. This dramatic woodcut appeared in the October 10, 1874, issue of *Frank Leslie's Illustrated Newspaper* following a New York City fire.

was a foregone conclusion. A flurry of contests that pitted the traditional hand pumpers against their new-fangled rivals quickly demonstrated why. Even when the hand-operated machines succeeded in disgorging longer streams of water (as happened during a face-off in New York City in 1856), the manual laborers invariably suffered fatigue while the steamers pumped on effortlessly.[23] By 1866, fifteen American cities had adopted steam. Just ten years later, 275 fire departments were routinely using "fire to fight fire" and, in the process, had created a new breed of "fireman" whose job it was to light the boiler fires. A spectacular display of American steam engines in action at the 1876 Centennial Exhibition in Philadelphia served as a fitting tribute to this revolution.

Coincident with modifications in fire engine design were numerous related innovations that aided fire protection in urban areas. Of fundamental importance was the introduction of hose, which not only enabled firefighters to work effectively while remaining a safe distance from burning buildings but also (in the case of suction hose) eliminated the tedious task

Telegraphic fire alarms, generally introduced in the 1850s, aided in the rapid response to fires. A hand-cranked electrical generator (below) triggered a bell or buzzer at the fire station (above). Courtesy of Chuck Deluca, Maritime Auctions, York, Maine.

of hand filling the pumper tubs. The hose itself was also improved over the years as inventors gradually replaced sewn leather (characteristic of the eighteenth century) with more durable riveted leather (invented in 1811) and ultimately rubber.

To quicken their response to fire emergencies, firefighters adopted a variety of other devices, including electric alarm systems (introduced experimentally in 1839 and perfected during the 1850s) and the fireman's pole (introduced in 1878). Horses, which had been used only sporadically to pull engines in the early days, became important members of the firefighting team in most places during the 1870s and 1880s, when the larger and heavier engines made manual hauling too difficult. As urban buildings became taller, firemen required additional equipment. Extension lad-

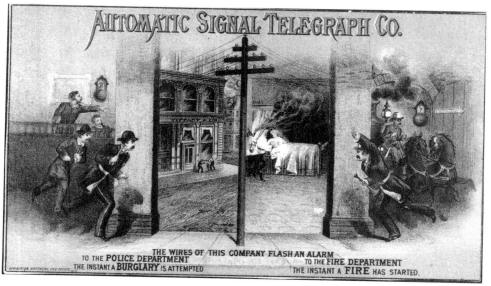

The telegraphic alarm of the Automatic Signal Telegraph Company served a dual role as police and fire alarm. Trade card from the Warshaw Collection, National Museum of American History, Smithsonian Institution Photo 89-13496.

ders (invented in 1861), water towers (introduced toward the end of the century), and a plethora of fire escape devices offered limited solutions to the problem of elevated fires. Chemical engines and sprinkler systems also appeared sporadically in various localities during the second half of the century.

Personnel

And what of the firemen who were in charge of these increasingly complex and sophisticated devices? Viewed as a group, they changed radically over time as well. In the early colonial period, firefighting was everybody's business. When an alarm sounded, all adult male citizens were obligated to respond, just as they did to militia musters. This arrangement, while haphazard, was not necessarily chaotic. In many large municipalities fire wardens, who were clearly identified by badges or other insignia, directed the firefighting efforts. New Amsterdam's Peter Stuyvesant imposed additional control by inaugurating an official fire watch (the famous Rattle Watch, so called in reference to their wooden alarms) in 1646. It is also generally assumed that when Bostonians imported their first fire engine in 1679 they must have created a company of men trained in its use.

Nevertheless, firefighting procedures were far from ideal. By 1715/1716, an irate resident of Boston felt compelled to tackle the problem in print. In his view, the woeful inadequacy of local efforts was directly attributable to the "great Crowd and Confusion that often falls out at a Fire; where no Body Governs, no Body will Obey, very few will Work, and a great many Lookers on . . . only incumber the Ground."[24] A short time later, a group of citizens took steps to improve the situation—at least so far as their own particular pieces of ground went. Uniting in 1717 in a mutual protection association, they pledged to protect fellow members' property both by reporting fires and by hastening to such blazes with buckets and bags in hand. (In the days when extinguishing a fire was a sometime thing, linen salvage bags were almost as important as water buckets.) Vigilance against looters was also a primary tenet in the Boston Fire Society's bylaws.

Benjamin Franklin knew a good idea when he heard one. Philadelphians, he had noticed, were pitifully lacking in "Order and Method" at their own fires. Thus, using the Boston society as a model, he organized the Union Fire Company in 1736. Franklin's rules required members not only to carry the standard bags and buckets to fires but also to attend monthly meetings to discuss their firefighting skills. If not the first volunteer fire company in America, as sometimes claimed, Franklin's group was certainly the inspiration for the many volunteer companies to follow.

And follow they did. By the time of the Revolution there were twenty-two fire companies in Philadelphia alone. Other organizations appeared elsewhere in the colonies—from Alexandria, Virginia, to Lancaster, Pennsylvania, to Brooklyn, New York. The movement expanded exuberantly during the early Federal period, and by 1835 the volunteer fireman had become a familiar urban figure throughout the East.

The advantages of this system were many. Instead of relying on the population at large to extinguish fires, communities could assign the work to a group of specialists. Most companies regulated their internal affairs through formal constitutions and bylaws and attempted to improve their skills through mandatory weekly meetings. The fact that volunteers were often attracted more by the social benefits of company membership than by the job of firefighting was not without compensation. Group solidarity had many advantages during fire emergencies, and in the quiet periods between fires it facilitated endless fund-raising activities that benefited the community as a whole. Many a town received shiny new equipment as a result of money-making balls and other events sponsored by the companies or, less commonly, through outright donations by well-to-do loyal volunteers.

On the other hand, problems were apparent from the start. Although the volunteers generally enjoyed the support of their communities (typically, they received the money collected from hearth taxes and chimney

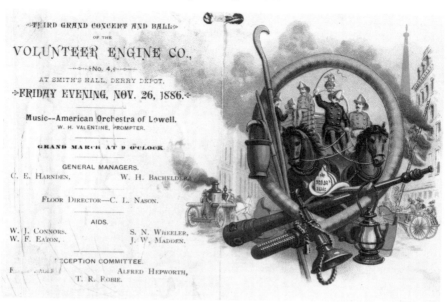

Volunteer firemen depended on public support. Their balls, often announced with beautifully engraved invitations like this one, were widespread in the last century.

fines and enjoyed exemptions from jury duty and militia service), local governments had little control over their firemen. Volunteer companies built their firehouses wherever they wanted, in the process overprotecting some neighborhoods while underprotecting others. They also promoted intercompany rivalries that interfered shamelessly with firefighting tasks. Stories of firemen fighting for possession of a hydrant while buildings burned to the ground are common relics of the antebellum years. The violence spilled over into nonemergency situations as well. In 1837, when a Boston fireman and an Irish immigrant had an argument, the ensuing riot resulted in several volunteer fire companies attacking an entire Irish neighborhood. With similar disregard for law and order, a Philadelphia gang called the Killers took over the Moyamensing Hose Company during the 1840s and used their power to intimidate the public. On one occasion, they set fire to a hotel and tavern owned by a black family and then fought off the fire companies that tried to extinguish the blaze.[25]

Baltimore's problems with its volunteers were so serious during the 1850s that the city had no difficulty identifying twenty-five different sorts of behavior that required immediate improvement. The diversity of the rules suggests the enormity of the problem:

Resolved, That no company, unless prepared with its apparatus to use the water of a pump, fire-plug, or other water source, shall retain it, if demanded by another company thus prepared. [Penalty: $20]

In returning from fires, or alarms of fires, the companies are requested to move with order and moderation; and should they meet another company, each is requested to take the right of the street. [Penalty: $20]

The use of intoxicating liquors among the fire companies composing this department [shall] be entirely discontinued. [Penalty: $5]

Expelled . . . members shall not be admitted into any other fire company, within twelve months from the date of expulsion. [Penalty: $50]

In all cases when a riot or disturbance takes place, the *company* with whom it commences, shall be held responsible, whether the act was committed by its members or not.

If any company shall be detected with lost hose in their possession, and neglecting or refusing to send it home after they shall have received two weeks' notice from the company losing the same, they shall be expelled from this department.

However valuable as a fireman a member be, if he is habitually turbulent and quarrelsome, it were better to dispense with his services as a fireman than that his continuance should be the cause of constant broils and disturbances. . . . it is hereby recommended that every such member be stricken from the roll of every company.[26]

This last provision appealed to officials in Cincinnati. Exhausted by years of conflict with undisciplined firemen, authorities in that city moved in 1853 to strike all volunteer firemen from the rolls and replace them with paid employees under direct municipal control. This radical move would have been unthinkable in the days of the labor-intensive hand pumper, but Cincinnati had made the transition to steam. With that pivotal decision, the city had some hope of instituting an orderly corps of professional firefighters. Albany, Buffalo, Providence, Portland, St. Louis, Baltimore, and Chicago were among the many cities that followed Cincinnati's lead and inaugurated paid departments during the 1850s. Boston eliminated its volunteer system in 1860; Philadelphia did so in 1871. With the unveiling of its extensive reorganization plan in 1865, New York City created the first full-time paid fire department in the country.

Pitfalls and Progress

This, in simplified form, is the standard story of firefighting in America. But while the major elements of the tale—the appearance of the early companies, the emphasis on the introduction of increasingly complex apparatus, and the depiction of slow but steady progress toward professional service—are correct as far as they go, they do not go far enough to provide a complete picture. In presenting a national overview based primarily on urban developments, such a summary necessarily sacrifices local details— details that reveal an extremely disorderly approach to firefighting in many communities throughout the nineteenth century.

In cities and towns alike, the acquisition of new apparatus was rarely a simple matter of choosing state-of-the-art tools. It was not just that volunteers frequently resisted innovation, although that was certainly part of the problem. Hose, horses, and steam engines all came under attack during the antebellum period because they were perceived to tarnish the manly image of the profession. More to the point, however, many communities found it financially impossible to purchase the latest equipment. Small towns across the nation limped along for years with only piecemeal protection: Howell, Michigan, platted in 1835, still relied on a bucket brigade in 1860; Greenfield, Indiana, incorporated in 1850, had a single hand-pulled hook-and-ladder wagon in 1880; Laurel, Delaware, had a hand pumper as late as 1899.

In many places, of course, there was no hope of introducing advanced engines because there was no adequate water supply. The model waterworks established in Bethlehem, Pennsylvania, in the eighteenth century was the exception. Most cities relied on cisterns, wells, and river or pond water for their engines well into the nineteenth century. Even then, the introduction of pressurized pipe water was slow.[27] More than eighty American cities had installed public water systems by 1850; but in thousands of small communities, particularly in the drier portions of the country, the delay was much, much longer. Tombstone, Arizona, was typical of western communities that used barrels of imported water for firefighting and other needs into the 1880s.

And if firefighting equipment appeared only haphazardly, so did firefighting personnel. Early fire companies have generated great popular interest, but the emphasis is somewhat misleading. Many small towns remained without organized fire protection decades after their founding. Stratford, Connecticut, though founded as early as 1639, did not establish a formal fire department until 1875. In Palestine, Indiana, which was formally laid out in 1838, citizens allowed more than half a century to elapse

VIGILANT NO. 1. FIRE DEPARTMENT, YORK, PA.

Financial constraints affected the type and age of
firefighting equipment in most communities. Fire
companies often documented their new buildings
and latest engines in photographs and postcards.

before they formed a department. The *Michigan State Gazetteer* for 1863
reveals that of the 743 townships listed, only 16 reported active fire departments. Some of those towns were undoubtedly nothing more than cross-
roads graced by a post office, but many were established communities
boasting Masonic societies, bands, and churches. By choice or by circum-
stance, they lacked organized fire protection. If these Michigan towns were
representative of communities elsewhere, they would continue to get by
without firemen until some terrible conflagration convinced them of their
vulnerability.

Even in communities where fire companies were on call, the likelihood
that those outfits were composed of well-trained professionals was remote.

Although paid departments had become the rule in large cities by the third quarter of the nineteenth century, volunteer firemen persisted—indeed, thrived—elsewhere. They offered a popular and economical solution to the problem of intermittent fire, but serious performance problems cropped up almost everywhere. William Dean Howells speaks of the "magnificent failures" that were the two fire engines in Hamilton, Ohio. "There seemed to be something always the matter with them, so that they would not work if there was a fire," Howells recalled.[28] Even when the machines worked properly, the men frequently did not. The Fargo, North Dakota, *Argus* reported more confusion than cooperation at a huge fire in that town in 1880: "A large crowd soon gathered and the cry was 'Where's the axe,' 'is there no rope,' 'where is your hook and ladder,' and 'what can we do.' Added to this was the noise of the switch engine which ran down in front of the fire and screeched. Unearthly yells added to the confusion."[29] One hundred and fifty years after Benjamin Franklin's sage observations, communities still sought "Order and Method."

With professionalism so lacking, it is perhaps not surprising that the public at large maintained an active role in firefighting throughout the nineteenth century. City departments may have discouraged casual interference in their efforts by the 1870s, but they continued to rely on their "buffs," a cadre of supportive friends and neighbors who rendered assistance by guarding the firehouse, closing its doors, and helping with the animals.[30] Small-town volunteers expected much more active aid in the form of hauling equipment and beating out flames. In rural areas, where organized protection was virtually nonexistent, the public was pretty much on its own. When prairie fires swept across the dry grasses, homesteaders had to dig their own trenches, build backfires, and soak down their outbuildings. William Allen White helped save his schoolhouse from such a blaze near Eldorado, Kansas. "We boys—little and big—took our coats off, soaked them in tubs of water from the school well and flopped out the fire, standing with the men of the town," he wrote with pride.[31]

Familiarity with various firefighting techniques was a good idea for everyone. "Have two pails filled with water in the kitchen where they will not freeze," counseled one self-help publication in 1869.[32] Extinguish chimney fires by throwing water on the hearth and then covering the opening with a thick piece of carpet soaked in water, advised another a few years later.[33] The acquisition of fire "annihilators," either store-bought or homemade, was yet another popular suggestion of the period. Evidently, if people were going to build and tend their own fires, it was assumed that they would acquire some knowledge of how to extinguish them too.

"Bedchamber Fire." Do-it-yourself firefighting was often the only way to get the job done, as this 1832 watercolor by Sophie Du Pont indicates. Courtesy of the Nemours Foundation, Wilmington, Delaware.

INSURANCE

Fire insurance was not an option for America's earliest settlers. In London such protection was almost unheard of until after the Great Fire of 1666, and it took longer still for the industry to become established in the colonies. When the first American fire insurance company finally appeared in Charleston in 1735, it had a long name—The Friendly Society for the Mutual Insurance of Houses Against Fire—and a short history: it perished in 1740 as a result of a raging fire that wiped out half of South Carolina's premier city. Nevertheless, the fire insurance business had gained a foothold in America, and during the second half of the century small-scale expansion was all but inevitable. Benjamin Franklin's Philadelphia Contributionship, established in that city in 1752, is a fondly remembered milestone in this process; but far more important in the long run was the emergence in 1792 of the nation's first stock insurance company: the Insurance Company of North America. By selling stock to help finance its operations, the company achieved greater capitalization than had been possible under the earlier mutual plans and so not only ensured its own success (it is still in business today) but also hinted at what the future might bring, namely, nationwide fire coverage that was wisely administered and financially secure.

By the last quarter of the nineteenth century many Americans held fire insurance, and companies had improved their reputations for stability and prompt payment. This advertisement from the *Saturday Evening Post*, December 4, 1926, combines statistics and drama to sell fire insurance.

That rosy vision did not materialize, however. To be sure, the nineteenth century saw a vast increase in the number of companies offering fire protection and in the quantity of their resources. By 1820 there were at least twenty-eight American stock companies doing business in six eastern states; and although many of these ventures failed because of the New York fire of 1835, the industry as a whole survived this setback and continued to expand. By the time of the Civil War, fire insurance had moved west and had become an aggressive multimillion dollar business. "Cash Capital and Surplus, $2,000,000," boasted the Aetna Insurance Company in 1863. "Losses Paid in Forty-two Years, $15,000,000."[34] But despite the vitality of the industry, prudent and conservative management gave way to the highly competitive and independent business tactics of the era. As a result, American fire insurance on the whole took on some of the characteristics of a circus sideshow. There were winners and losers, hawkers and hucksters, and, of course, amply stacked decks in favor of the management.

Both the industry and the public contributed to this topsy-turvy situation. Consumers wanted low premium rates and frequently awarded their business to cut-rate vendors with inadequate resources. Naturally, when fire struck there were plenty of disappointed policyholders. Furthermore, while demanding protection from fire loss, the public willfully ignored even the most basic fire safety tips in everyday building and living habits. Chicago's continuing indifference to fire-resistant construction after the great fire so upset the city's insurers that they united in 1874 and threatened to cancel all insurance policies unless remedial action was taken.

On the other hand, insurance companies had plenty of faults of their own. They were known to fix prices at levels that far exceeded a fair profit margin and then, adding insult to injury, to balk at paying claims. Household handbooks of the second half of the nineteenth century are full of cautionary notes concerning these sorts of problems. Hamilton's *Household Cyclopedia* (1875), for instance, warned readers to avoid altering insured premises in ways that might negate their policies and to refrain from moving to new dwellings with the expectation that the old policies would remain in effect. The *Universal Self-Instructor* (1883) carefully explained that if goods were moved from one place to another, notice must be given immediately to the company "and the policy altered in accordance with the new state of facts." The *American Domestic Cyclopedia* (1890) offered more general advice. Given the many problems the public had with fire underwriters, this manual concluded that fire prevention was the best insurance of all. "Have your fire safe!" it proclaimed in 1890. "Insurance money is hard enough to get when all the conditions have been justly complied with. It is terrible to burn out!"[35]

No one on either side would have disagreed with that. And because it was in everybody's interest to have a stable fire insurance industry, numerous efforts were made throughout the century to bring consumer and vendor closer together. Government regulation, which began in 1837 when Massachusetts required companies to back their policies with adequate funding, was ultimately accomplished in every state. The industry also made an effort at self-regulation with the creation of the National Board of Fire Underwriters in 1866. Although the establishment of uniform rate scales was part of its mission, this voluntary association also tried to solve some of the problems that hurt everyone: arson, shoddy building practices, and unenforced building codes. These efforts were supplemented during the 1880s by a number of regional trade associations with similar goals.

The strongest regulator of all, however, was fire itself. In the wake of every major disaster, weak companies were eliminated while the survivors gained strength. The winners also acquired vastly improved risk assessment techniques, which stood them in good stead for future transactions. After the Chicago fire, for instance, some fifty fire insurance companies failed, and even more succumbed after Boston's holocaust. By the time of the San Francisco blaze of 1906, however, the industry had begun to achieve greater stability. A number of bankruptcies occurred in the wake of this disaster, but most analysts agree that the industry had sharpened its expertise at calling the odds.[36]

Risk assessment was also an integral part of the game for members of the public, who had to decide whether or not to invest in some form of fire protection for their homes and businesses. Many chose not to do so and "lost their all" when fire struck. For these victims, an informal support network made up of family, friends, and the community at large often provided limited financial assistance. In the mining town of Aldridge, Montana, for example, the quick response of local business and music organizations was typical of communities across the country at the turn of the century:

> Mr. Hynd was filling a can with coal oil and spilled some on the floor. It being dark he thought that it was only a small amount so he put a match to it and both him and Mrs. Hynd had a very narrow escape. As it was, Mr. Hynd was burned very severely about the face. They did not save any of their household furnishings or wearing apparel and their loss is about $500. A joint soliciting committee was appointed by the Miners Union and AOUW, and they have collected up to the present time $250 to give them a fresh start in the world. Messrs. Herst Beever, Ed Wright and William North have organized the Aldridge orchestra and gave a dance in the Union Hall last Wednesday for the benefit of Alex Hynd.[37]

By contrast, many individuals bought fire insurance and had reason to rejoice in that decision. Even if they were underinsured (a common problem), they were usually able to collect something with which to begin life anew, and some people did considerably better than the bare minimum. The story of Adolph H. J. Sutro demonstrates the possibilities, if not the most common results. Sutro had been running a lucrative smelting operation near Virginia City, Nevada, when his smelter suddenly burned down in 1863. Well aware of the risks of such a venture, he had insured the business adequately and was able to collect. Rather than rebuild the business, however, Sutro took off in a new direction. He was convinced that silver mining operations would be greatly enhanced if the miners could get at the ore from below as well as above. Such a scheme required extensive tunneling, and therein lay Sutro's dream. Using his insurance money to get the project started, Sutro worked tirelessly to oversee the building of a tunnel that, when finally completed in 1878, measured twelve feet in width and an incredible four miles in length. Two years after completion of the tunnel, Sutro sold out, went to San Francisco, and made millions of dollars in real estate.[38]

OTHER RESPONSES

Fire prevention, firefighting, and fire insurance represent the most visible and easily documented solutions to America's fire problems in the pre-electric era. They were not the only responses, however. Two additional weapons enhanced the country's firefighting arsenal, and although they were attitudes of mind rather blueprints for action, they were effective in their own unique ways. At times, in fact, these psychological defenses were so powerful that they functioned as virtual safety nets in times of crisis. When more pragmatic measures failed, these responses were able to fill the void and help Americans tolerate a world beset by the ravages of the flame.

The Fireman Mystique

It was probably inevitable that a society that regarded fire as more than a chemical reaction would perceive firemen as more than community workers with a job to do. Even so, the notoriety achieved by American firefighters during the last century is astounding. They were, as the British journalist Charles Mackay claimed, a *"great* institution,"[39] and the fact that they often failed to protect the public from destruction by fire scarcely mattered. As a group, they gave the impression that they had extraordinary powers, and that impression, illusion though it might have been, helped to bolster the public's confidence that they might, eventually, gain control over the destructive aspects of fire.

Not surprisingly, firemen themselves contributed to the development of this image. Believing themselves to be larger than life, they devised endless in-house rituals that had nothing to do with firefighting and everything to do with exalting the fraternity. Since membership was selective (frequently old members chose the new), admission to the group was usually commemorated with lavishly engraved certificates and ornate company badges, belts, and buckles. To recognize special occasions such as retirements and important anniversaries, fire companies commissioned a vast array of additional mementos. These ranged from handsomely engraved

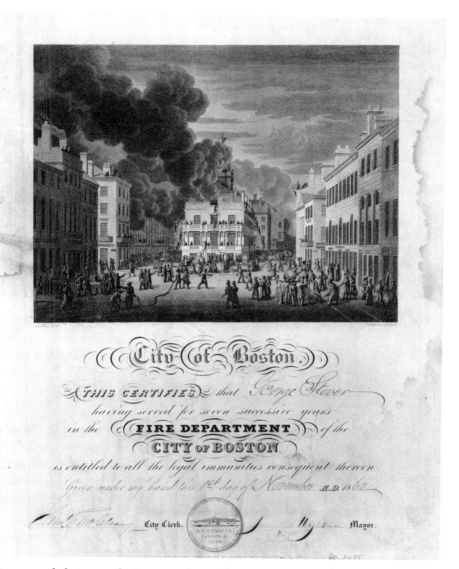

Firemen and their contributions were honored in many ways. In the 1860s, elaborately engraved certificates were awarded for seven years of service in the Boston Fire Department. Peters Collection, National Museum of American History, Smithsonian Institution Photo 60446-F.

silver speaking trumpets and punch sets to finely crafted presentation helmets, water pitchers, neckerchief rings, and shields.[40] Company ceremonies were as common as the company pet.[41] The acquisition of a new engine, for instance, traditionally entailed a baptismal washing or an inaugural demonstration. And when a firefighter died, his fellow firemen not only participated in the funeral rites but often deployed the company hose cart for use as a hearse.

Bolstered by these and other customs, firemen developed a fierce loyalty to their comrades. Even so, they never forgot that they were public servants, and—with unapologetic emphasis on the word *public*—they exhibited themselves as often as possible to the local populace. The parade was their favored forum; indeed, by the end of the nineteenth century it would have been unthinkable for a town to have a procession without them. Firemen took to the streets on such special occasions as the opening of the Erie Canal, Andrew Jackson's funeral, and the dedication of the Statue of Liberty, as well as on countless annual holidays. They also organized their own processions, frequently inviting neighboring companies to join in the festivities. On one memorable occasion in the autumn of 1850, nine hundred firemen from the Midwest gathered with fourteen engines in tow to parade on the streets of Chicago. For the great Lancaster, Pennsylvania, parade of 1867 local firemen rejoiced in the participation of more than twenty visiting companies. According to the local press, the ensuing procession was not just "magnificent"; it was quite simply "The Biggest Thing of the Kind Ever Got Up in the State Outside of Philadelphia."[42]

For their public appearances, firemen invariably put on an elegant show. The apparatus was not so much maintained as it was glorified. Oil paintings of eagles, American flags, burning buildings, and semiclad women decorated the side panels of the early hand pumpers. Sometimes these images were the work of local amateurs, but more often fire companies hired professional sign painters and even such well-known artists as John Vanderlyn, Thomas Sully, John A. Woodside, and John Quidor. The artistry that appeared on the later steam engines was less pictorial but no less elaborate. Bright red paint, floral figures in gold leaf, and geometric patterns on the wheel spokes were common decorative features for many turn-of-the-century vehicles.

The firemen's clothing was as handsome as their equipment. In 1840 the Protection Engine Company of New York introduced a practical, but attractive, working uniform that eventually became standard throughout the country: a double-breasted red shirt, a black tie, black trousers, and boots. The now-classic fireman's helmet with long back brim had been introduced in 1828 by Henry Gratacap, and this too was widely adopted by mid-century.

Elaborately engraved silver speaking trumpets were prized trophies
of the firefighter's profession. Courtesy of Chuck Deluca,
Maritime Auctions, York, Maine.

Marching was much too important a pastime, however, to allow uniform design to stop with these basics. Most companies maintained two sets
of uniforms, one for firefighting and one for ceremonial purposes. For the
latter, only imagination and the pocketbook put a limit on the number and
intricacy of decorative accessories. There were parade coats, capes, gloves,
belts, ties, gauntlets, and shields. Most prized of all were decorative parade
hats, usually constructed from felt or leather and emblazoned with the
company name and a complementary picture. For the Lafayette Hose
Company of Philadelphia, a portrait of Lafayette was the inevitable choice
of adornment for their stovepipe parade hats. Blessed with a less specific
name, the Hope Hose Company had the freedom—one might even say the
audacity—to place the image of a statuesque woman on their helmets. Like
the fire engine artists, the hat decorators could be amateurs or local craftsmen, but in at least one case a professional was called in. Writing to his son
John in 1894, Philadelphia fireman Thomas Peto asked a favor of the
noted trompe l'oeil specialist: "If I send the front of my fire hat out to you
will you paint a picture of a hand engine on it?"[43]

Parading apparel was supplemented by a plethora of specially designed
parade equipment. Most common were ornamental axes and speaking
trumpets, but some units, primarily hose companies, undertook the extraordinary expense of maintaining stunning parade vehicles. One such

This handsome parade vehicle was built by Joseph Pine and belonged to the Rapid Hose Company of Kingston, New York. This view shows its appearance around 1890. Courtesy of the Rapid Hose Company.

carriage was purchased by the Rapid Hose Company of Kingston, New York, in the 1870s. Built by Joseph Pine of New York City, the vehicle boasted a nonfunctional hose reel decorated with velvet and cut glass mirrors. The impression was pleasing but not extraordinary. Over the next ten or twelve years, therefore, the Kingston volunteers refurbished and improved their carriage. When it was finally finished around 1890, its features were legendary: a silver-plated iron frame, a delicate ornamental railing, silver-plated bells, gold and silver candle lamps, an oil painting, and a large cast statue surmounting the whole.[44]

With accoutrements such as these, parading firefighters could hardly fail to capture public interest. But showmanship among the volunteers ran deep, and most companies devised opportunities to demonstrate their skills in public. Such an exhibition was successfully enacted in 1824 as part of Lafayette's welcoming parade in New York City. On this occasion the firemen of Brooklyn and Manhattan arranged three ladders in a tripod and placed a burning torch at the pinnacle. Then, in an elegant and skillful display, the last two engines in the procession activated their pumps and extinguished the flames.

Other companies preferred to emphasize strength and endurance rather than finesse. Such a goal was easily achieved by staging pumping contests and engine races. As befits a great metropolis, New York's firemen put on a magnificently vigorous show for their "Old New York Fire Department Celebration" in 1885. On this occasion the men not only raced through the city streets with their engines in tow, but they also set a building on fire and then promptly put it out. Small towns had fewer resources but no less enthusiasm. Practice sessions were often designed to draw a crowd, and in many places hook-and-ladder races organized by the fire department became cherished Independence Day events. As an adjunct to these local activities, most companies also flexed their muscles from time to time for their fellow firemen by participating in innumerable invitational tournaments, competitions, and conventions.

If all of this sounds a bit self-serving, perhaps it was. And yet, firemen did not have to create the fireman mystique all by themselves; there were always plenty of enthusiastic supporters outside the fraternity to help. This reverential attitude came easily to William Dean Howells while he was growing up in Hamilton, Ohio:

> My boy would have liked to speak to a fireman, but he never dared; and the foreman of the Neptune, which was the larger and feebler of the engines, was a figure of such worshipful splendor in his eyes that he felt as if he could not be just a common human being. He was a store-keeper, to begin with, and he was tall and slim, and his black trousers fitted him like a glove; he had a patent-leather helmet, and a brass speaking-trumpet, and he gave all his orders through this. It did not make any difference how close he was to the men, he shouted everything through the trumpet; and when they manned the breaks and began to pump, he roared at them, "Down on her, down on her, boys!" so that you would have thought the Neptune could put out the world if it was burning up.[45]

These childhood sentiments were not altogether alien to the adult psyche. Of the grand firemen's parade held in Lancaster in 1867, the local press exclaimed, "The universal expression of opinion is that they are physically the finest set of men ever seen together by our people. Their conduct throughout was admirable and they leave behind them a reputation for gentlemanly demeanor of which they may well be proud."[46] Alfred M. Downes, who wrote a glowing tribute to New York's firemen in 1907, lavished praise on American firemen everywhere. According to

SAMPLE FROM
HUDSON VALLEY PAPER CO.,
520 AND 522 BROADWAY,
ALBANY, N. Y.

INVITATION CARD.
No. 1599.

Firefighting themes graced all kinds of everyday items. A brave fireman in a colorful setting made a distinctive invitation circa 1870.

THE LIFE OF A FIREMAN.

Currier and Ives paid tribute to firefighters in their popular series of lithographs entitled "The Life of a Fireman." Peters Collection, National Museum of American History, Smithsonian Institution Photo 60477-C.

The role of firemen is glorified in this fanciful lithograph: a company of firefighters helps to lay the transatlantic cable. Peters Collection, National Museum of American History, Smithsonian Institution Photo 49746-D.

Downes, "One and all, enginemen and hook-and-ladder men, they stand shoulder to shoulder fighting fiercely and fearlessly to subdue the flames and protect property and to save human life."[47]

Physical evidence of this boosterism appeared almost everywhere. In the years before the Civil War firefighters and their equipment adorned tea sets, band boxes, and weathervanes, as well as endless lithographic prints.[48] Later in the century, firemen frequently ornamented such items as decanters and invitation cards, and they were even featured as heroes in the first American movie to have a plot, Edwin S. Porter's *The Life of an American Fireman* (1903). When A. Weingartner produced a lithograph to commemorate the laying of the Atlantic cable in 1858, he decided that the most effective way to show that Americans were an energetic and boisterous lot was to dress them as fire laddies. And when the great American bandleader Patrick Gilmore staged his National Peace Jubilee in Boston in 1869 he, too, paid homage to firemen. Amassing one hundred red-shirted

Patrick Gilmore assembled one hundred red-shirted firemen to form a mighty percussion section for Verdi's "Anvil Chorus" in 1869 and again in 1872. This is the 1872 gathering. Photograph courtesy of Paul A. Munier.

firemen from the Boston department, Gilmore presented each of them with an anvil and a hammer. Then, with a flick of the baton, the maestro transformed the firemen into a hearty percussion section for an outstanding performance of Verdi's "Anvil Chorus."[49]

Although Gilmore's performance was somewhat unorthodox, bandsmen had always had a special affection for firemen. This is not surprising, since in some towns the memberships overlapped and almost everywhere a commitment to good fellowship characterized both the volunteer fire department and the amateur town band. In a sense, they worked for each other. Brass bands furnished the music for firemen's parades, balls, and relief fund benefits. In return, the firemen not only extinguished fires for bandsmen but also supplied rousing themes for band music. There were firemen's quadrilles and firemen's polkas, as well as more serious dedicatory pieces, such as Francis Johnson's "Philadelphia Firemen's Anniversary Parade March." Even more popular were the many programmatic pieces that wove the sounds of fire bells and shouts of "Fire!" into the melodic lines. One such work, written by bandleader D. W. Reeves in 1890, traces in sound the entire story of a night alarm: the call to a fire, the race of the horses and engines to the blaze, the battle with the flames, and the peaceful return to the engine house.

Taken together, these songs, pictures, and other cultural artifacts provide irrefutable proof that firemen held a conspicuous position in America

Brass bands and cheering crowds marked gatherings of firefighters, such as the June 1895 assembly of these Portland, Maine, veterans. Photograph by J. I. Barbour, Portland.

during the last century. Their prominence often went far beyond acts of firefighting and celebration, however. In many towns, in fact, firemen took on important civic roles. They coordinated a wide range of social events in communities across the country, and on occasion they even achieved some political status. This is hardly surprising, considering that prominent citizens routinely joined their local companies (George Washington, Benjamin Franklin, Robert Fulton, James Buchanan, and Ralph Waldo Emerson all served in this capacity). The prestige of fire companies was such, however, that the reverse was also possible. That is, a relatively unknown individual might use membership in a fire company as a stepping-stone to political power. William Marcy ("Boss") Tweed is probably the most notorious person to take this route, but there were plenty of others. In times of social upheaval, entire companies are known to have done so. During the American Revolution, for instance, the firemen of Charleston resisted the Stamp Act in the style of the Sons of Liberty, and during the

Fire themes provided inspiration for composers and attractive covers for sheet music publishers, such as this example from 1907.

Civil War a number of fire companies—notably Colonel Elmer Ellsworth's Fire-Brigade Zouaves—enlisted for military service. Later, when military associations were banned as part of Reconstruction politics, southern whites joined fire companies instead and gradually transformed these innocuous associations into grass-roots military units dedicated to white supremacy.

These activities suggest that there was a dark side to the fireman mystique, and, indeed, other historical evidence bears this out. Satirical prints that lampoon ineffective firemen, scathing editorials that complain about the rowdiness of the local volunteers, and bawdy songs that mock the brotherhood's propensity for drunkenness—these sorts of documents were printed and distributed alongside the glowing testimonials. Curiously, though, the public at large remained largely undisturbed—or, at most, temporarily distracted—by such shortcomings on the part of volunteers. Although communities certainly tried to reduce some of the most flagrant excesses, they also seemed willing to tolerate the fact that firemen, like the fires they fought, were rugged and somewhat erratic.

In the end, it was this attitude that gave rise to what is perhaps the most creative element in the evolution of the American firemen mystique: the character of "Mose, the Bowery B'hoy." Although Mose is based on a real-life New York City fireman named Moses Humphreys, Mose the folk hero was catapulted into the national consciousness by way of the theater. He first appeared in public on February 15, 1848, in a melodrama called *A Glance at New York*. The creation of actor Francis S. Chanfrau in association with playwright Benjamin A. Baker, this play introduced Mose as a boastful, brawling Bowery boy who loved a good fight but loved his fire engine most of all. The play was largely episodic and included a number of scenes in which Mose saved a greenhorn from city slickers. But plot was clearly secondary to personality. The real appeal of the production was the irresistible Mose, who swaggered about in his red shirt and stovepipe hat and engaged in such memorable firehouse repartee as "Get off dem hose or I'll hit yer wid a spanner."[50]

The play was an instant success. After seventy-four performances in New York, traveling companies took it on the road. By then the American public was hooked on Mose. The lusty firefighter appeared on lithographs and posters and in any number of newspapers and joke books. He was a legend among American schoolboys, who imitated his speech, and his larger-than-life personality gradually insinuated itself into ballet and circus performances. Mose's theatrical career flourished as well. During the 1850s there were several successful Mose sequels, including *New York as It Is!*, *The Mysteries and Miseries of New York*, *Mose in California!*, and *Linda*,

Mose, fireman extraordinaire, was the subject of plays, lithographs, and fireside stories. This 1848 New York lithograph depicts Mose's creator, actor Francis S. Chanfrau, in the role. Peters Collection, National Museum of American History, Smithsonian Institution Photo 60446-B.

the Cigar Girl; or Mose Among the Conspirators. Some of these dramas featured Mose and his cohorts extinguishing fires and saving babies from the flames; all glorified the antics of the irrepressible Bowery boy. Although enthusiasm for dramas about Mose began to diminish by the 1860s, the character lived on for decades as the hero of orally transmitted tall tales. By the beginning of the twentieth century the valiant fire laddie had taken on superhuman characteristics, and it was not uncommon to hear him described as standing twelve feet tall and smoking two-foot cigars.

As folklorist Richard M. Dorson has pointed out, Mose is a true American folk hero, on a par with Davy Crockett, Mike Fink, Sam Patch, and Yankee Jonathan. Like them, he was strong, irreverent, and a spokesman for the common man. Unlike these better known characters, however, Mose was a city man—a cigar-chomping, beer-swilling, street-smart dandy.

Perhaps it says something about the diminished role of fire in contemporary life that we have largely forgotten Mose today. It surely says something about the centrality of fire in the nineteenth century that our forebears, in creating America's first urban folk hero, chose to make him a fireman.

The Phoenix Response

There was yet another psychological weapon deployed against fire in early America. For want of a better name, it might be called the "phoenix response," in honor of the legendary bird reborn from its own ashes. Less a call for specific action than a mode of thought, this philosophy interpreted fire not as a tragedy but as an opportunity for improvement. Its advocates, united only by their intensely resolute thinking, perceived smoldering ruins not as scenes of desolation but as fertile seedbeds for new growth. With brass band–style bravado, they sounded the peacetime rallying cry to "rebuild," "reinvest," and "reform." Persistently, almost relentlessly, they counseled optimism in the face of disaster.

In a sense, this response harks back to prehistoric times, when early humans captured the devastating power of forest fires and reprogrammed it for their own purposes. It is more directly related, however, to various religious and philosophical theories that in modern times have tried to ascribe some higher purpose to human suffering. During the eighteenth century, for instance, fire sermons, in which "the sins which provoke the Lord to kindle fires are enquired into," were common in New England.[51] By the early nineteenth century, intellectuals and writers such as James Fenimore Cooper had recast the Puritan sinner into the Yankee materialist and had begun to portray fire as "a fearful admonition for those who set their hearts on riches."[52] In *The House of the Seven Gables*, Nathaniel Hawthorne pays tribute to both traditions through the character of Holgrave, who says of the old Pyncheon house, which clearly had been built on a shaky foundation of greed and corruption, that it "ought to be purified with fire,—purified till only its ashes remain!"[53]

Purification figures only minimally in the phoenix response, for the essence of this philosophy is not based on pessimistic, dark, or moralistic views. Rather than focusing on past failures and shortcomings, this outlook concentrates on the future. And instead of peering inward toward the realm of the spirit, it unapologetically looks outward to the material world and to the manifold possibilities for social reform. For these reasons, the phoenix response was particularly well suited to public pronouncements in the wake of serious municipal fires. Indeed, Chicago adopted the phoenix as its official symbol at the time of the 1871 fire, and such imagery was perpetuated in countless editorials thereafter. Joseph Medill caught the essence of this boosterism in a column published in the *Chicago Tribune* in 1871: "Looking upon the ashes of 30 years' accumulations, the people of this once beautiful city have resolved that CHICAGO SHALL RISE AGAIN."[54] And so it did. Chicagoans undertook rebuilding with such enthusiasm, in fact, that they went on to construct the immense "White City" to host the 1893 World's Columbian Exposition. Not to be outdone, Baltimoreans talked about rebuilding their city even before the flames of the 1904 conflagration had died out; and San Franciscans, momentarily stunned by their losses in 1906, not only reconstructed their metropolis but also won the right to sponsor the 1915 World's Fair. Never mind that most projected improvements fell short of the euphoric expectations. The point was that there *were* great expectations.

Small communities found the phoenix philosophy as useful as these larger ones did. Thus, when a disastrous fire wiped out numerous shops and barns in Howell, Michigan, in 1860, the local editor had his rhetoric ready: "With characteristic energy . . . Mr. Melvin commenced a new building while the embers of the old were still burning, and two days had not elapsed ere the 'anvil chorus' was ringing in his new shop."[55] With similar optimism, officials at the University of Virginia chose to characterize the terrible fire of 1895 not as a calamity but as a "turning point" that afforded the university the opportunity to modernize the otherwise untouchable Jeffersonian plans.

Nowhere did the power of positive thinking work as well as in the world of business, where boosterism had almost always contributed to growth and profit. Throughout the nineteenth century, companies that suffered serious fires turned such temporary setbacks to their advantage by publicizing the fascinating particulars of their remarkable recoveries. Lyon and Healy, Chicago's great musical instrument supply house, mastered the technique in a brief company history first published in the *Chicago Evening Journal* in 1880 and subsequently reprinted in the company's widely circulated merchandise catalog. The company, which had been founded in 1864, prospered continually until September 4, 1870, when a fire destroyed the firm's headquarters and $100,000 worth of goods. "Nothing daunted" and "while the flames were yet devouring their goods," the owners began to rebuild.

The growth of the C. G. Conn band instrument company was barely slowed
by the destruction of its factory by fire in 1883. The rebuilding of the facility
just two months later provided Conn with an effective advertising ploy. Hazen
Collection, National Museum of American History, Smithsonian Institution
Photo 90-8370.

But, as fate would have it, the firm was no sooner open for business again
than it was caught in the citywide conflagration of 1871. Company losses
this time exceeded $150,000. As the history was quick to point out, how-
ever, "That appalling calamity, great as it was, did not by any means discourage Lyon & Healy. Scarcely had the smoke cleared away ere they were
to be found temporarily located in the modest little frame store . . . where
they remained for about four months." From then on, there was no stopping the company; and, as the narrative boasted in its conclusion, Lyon
and Healy soon became the "largest and most flourishing music business on
the continent."[56]

This was a self-congratulatory story, to be sure, but it was also a very
practical one. In the wake of a devastating fire, one could either give up or
go on, and there is little doubt that the American work ethic favored the
latter for public pronouncements. For individuals, of course, recouping
from fire loss was often easier said than done. P. T. Barnum, who diligently
rebuilt his American Museum after a fire in 1865, lacked the energy and
will to do so after another in 1868. As he wrote in his autobiography, "I
was not now long in making up my mind to follow Mr. Greeley's advice on
a former occasion, to 'take this fire as a notice to quit, and go a-fishing.'"[57]
Ralph Waldo Emerson, the sage of Concord, lacked the luxury of such a
choice. When his house burned in the summer of 1872, the shock of the
tragedy caused physical damage that seriously inhibited his ability to travel
and write.

P. T. Barnum suffered two disastrous fires at his American Museum in less than four years. The second blaze, in 1868, left him with no enthusiasm for rebuilding. Peters Collection, National Museum of American History, Smithsonian Institution Photo 55244.

Surely there were many others for whom the notion of rebuilding was equally meaningless. Nevertheless, as Kenneth Boulding has pointed out, images or metaphors are often more important to a culture than factual reality.[58] There is little doubt that Americans of the nineteenth century believed that the appropriate response to serious fire loss was to pick up the pieces and carry on. The entries for 1813 in the log book from the Martha Iron Furnace in southern New Jersey give a minimalist picture of this attitude:

> June 2[:] Great conflagration. The Furnace and Warehouse was this day entirely consumed, but fortunately no lives lost. John Craig got very much burnt.
> June 4[:] Ball & Richards arrived here this evening and concluded to build the Furnace up again.[59]

A more eloquent description of the attitude is furnished by Henryk Sienkiewicz, a writer who arrived in Chicago in 1876. Sienkiewicz encountered such vibrant scenes of recovery and rebirth in the great midwestern city that, despite his overall disappointment in American culture and society, he praised the optimism of the people: "Within a few years no traces of the fire will remain—and if the town should burn again, it will once more be rebuilt. It will be rebuilt twice or even ten times, for the energy of these people surmounts all misfortunes and all disasters."[60]

Sienkiewicz's assessment is a fitting tribute to what was clearly the country's most forward-looking method of "fighting back."

5

Keep the Home Fires Burning: Fire Tending and Maintenance

In the popular fiction of the day, nineteenth-century Americans spent a prodigious amount of time coaxing their fires to life. Characters "stirred" their fires and "fixed" them. They "laid" fires, only to have to "mend" them later—sometimes several times in a single scene. Although most skillful writers of the period were adept at exploiting the flame for metaphoric purposes, the utilitarian aspects of fire building form the ever-present back-drop for many tales. It is not by chance that one of Jo's trials in *Little Women* entails keeping the fire going (she fails) or that the first task accomplished by little Ellen Montgomery in her journey through *The Wide Wide World* is to "poke the fire" and toast a piece of bread to perfection.[1]

As diaries and autobiographical accounts make plain, fire maintenance was every bit as time-consuming in real life. "Had an offal time to get breakfast, the fire would not burn," complained Mary Bennett, who was late for school one winter morning in 1868 because of a recalcitrant fire. "I have seen my father work for an hour before he succeeded in getting a blaze," recalled another youth from the same era. For pioneers on the trail, the problems were compounded, though humor went a long way toward relieving the strain. "Although there is not much to cook," wrote Helen M. Carpenter as she headed west in a wagon in 1857, "the difficulty and inconvenience in doing it, amounts to a great deal—so by the time one has squatted around the fire and cooked bread and bacon, and made several dozen trips to and from the wagon—washed the dishes . . . and gotten things ready for an early breakfast, some of the others already have their night caps on."[2] In light of such testimony, it is easy to understand the attitude of Emma Allison, operator of a steam engine featured at the Women's Pavilion during the Centennial Exhibition of 1876. When questioned about her ability to run such sophisticated machinery without assistance, Allison claimed breezily that it was less tiring to run her six-horsepower Baxter engine than it was to cook over a kitchen stove.[3]

Energy production, in short, was a labor-intensive activity. As the majority of Americans of the last century knew, the only satisfactory way to compel fire to work for them was to do enormous amounts of work themselves. Virtually every stage in the process—from starting fires to feeding them to managing their power for maximum efficiency—required decisionmaking and the physical manipulation of tools and fuels.

FIRE TENDING: PROBLEMS AND PROCEDURES

Starting Fires

Oxygen and fuel are all around us. The missing ingredient for starting a fire is heat. Dry leaves, grass, paper, and other kindling materials ignite at about three hundred degrees centigrade, whereas common fuels such as charcoal and pine require even higher temperatures to burn. The trick in fire starting, therefore, is to convert some other form of energy into heat localized at the fuel such that the kindling will catch fire and ignite the underlying combustibles.

Advances in this field came slowly. American colonists of the seventeenth and eighteenth centuries started their fires by such age-old friction techniques as rubbing two pieces of wood together and striking flint with steel. The latter technique was by far the most common for Americans of European descent: the fire starter held a piece of steel over some tinder—typically charred linen and wooden sticks dipped in sulfur—and struck the metal with a piece of flint. If all worked properly, a little spark of steel would eventually fall onto the linen and ignite it. Then, assuming the linen was dry enough, it could probably be coaxed into smoldering long enough to ignite the end of the chemically treated "match," which in turn could be applied to the waiting fuel. Tinderboxes simplified the process somewhat by fastening the steel to a stable surface and protecting the tinder from adverse atmospheric conditions. Yet, even though a fire *could* be started with flint and steel in less than a minute, the inescapable truth was that it often took much longer. A survey conducted in Britain as late as 1832 revealed that domestic servants required anywhere from three to thirty minutes to produce a flame.[4]

The sheer difficulty of fire starting probably accounts for the reverence accorded perpetual flames in some cultures. Certainly most American householders preferred to keep existing fires going rather than start from scratch. To this end, they carefully banked their fires at night (a well-protected fire kept eight to twelve hours) and simply transferred pieces of the "old" fire to "new" ones as needed. If these measures failed, as they often did, it was considered easier to go and get a live coal from a neighbor than to put the tinderbox into action. Or, rather, adults considered it easier to send a child on such an errand. The historical record is full of

Blowing the Fire, 1842   ***Using a Match,*** 1842

Matches greatly simplified the process of starting a fire. Illustration from
Herbert Manchester, *The Diamond Match Company: A Century of Service,
of Progress, and of Growth, 1835–1935* (1935).

lively reminiscences of the common childhood experience known as "borrowing fire."[5] David Shryock, the son of a farmer, remembered,

> If you happened to get out of fire, you had to go to the neighbors for it. Well, one time we ran short and mother sent me down to Mr. Parker's after some. It was getting quite late in the evening. Got the fire and started home. By that time it was beginning to get dark and I was hustling along. It was through the woods, only a small place cleared away for a road. When I had gotten within sight of home, there was an awful yell, I thought it was right behind me. I dropped the fire and if ever a fellow did tall running I was the chap. Told mother there was some awful thing that yelled at me and scared me, so I dropped the fire. By the way, I didn't look back until I reached home. One of the girls went back and managed to get enough fire to start with.[6]

Such occurrences may have seemed exciting in retrospect, but they were serious deterrents to efficient work. For years, chemists sought improved methods of fire starting, and toward the end of the eighteenth century their efforts began to pay off. In 1781 the "Phosphoric Candle" or "Ethereal Match" appeared in France. This device, which relied on the fact that phosphorus combusts spontaneously in air, consisted of a strip of paper dipped in phosphorus and encased in a glass tube. As soon as the glass was broken, the paper burst into flame. The "Instantaneous Light Box" achieved the same effect using a different chemical reaction. Invented in 1805, this little kit contained a bottle of sulfuric acid and a number of wooden splints tipped with a mixture of potassium chlorate, sugar, and gum arabic. "One of these you put into the bottle, and pulled it out aflame," recalled Edward Everett Hale, an enthusiastic user of this novelty during his childhood.[7]

But novelty is exactly what such things were. The modern friction match evolved along different lines, relying, as the name implies, on the frictional application of heat to highly flammable chemicals affixed to the end of a stick. An English apothecary named John Walker pointed the way in 1827 by producing matchsticks tipped with a mixture of antimony sulfide and potassium chlorate; they ignited instantly upon being pulled through folded "glass paper." Three years after this debut, a French inventor substituted phosphorus as one of the key ingredients. In 1836, production of phosphorus friction matches began in the United States.

Although the breakthrough was a long time coming, the manufacture of friction matches was fairly straightforward. Popular publications of the 1840s and 1850s occasionally provided recipes for "Lucifer Matches" so that more adventuresome readers could make them at home. It was simpler, however, to buy matches ready-made, and by the middle of the nineteenth century the American public could choose from a variety of domestic and imported brands. "We warrant all our matches to be good, and to keep in any climate . . . *impossible* for them to ignite until unpacked," proclaimed the Boston Union Match Company. "Entirely free from Sulphur or any unpleasant odor and rendered waterproof by varnishing each match . . . superior to any others," boasted a dealer of the rival Strela Matches.[8] As David Shryock recalled, the high cost of matches meant that at first "poor folks could not afford to use them," but mass production soon solved that problem.[9] From the 1860s until about 1910, "match safes" loaded with commercially produced matches could be found in kitchens throughout the country.

The impact of this revolution was extraordinary. Textbooks ranked matches with the cotton gin and the steamboat in terms of the benefits bestowed on society. A contingent of radical Jacksonians even took on the appellation "Locofocos" when they used matches by that name to light up a meeting hall. On the other hand, this innovation did not come without a price. Not only did manufacturers exploit large numbers of children to sell their products on the streets, but rats, who apparently enjoyed the taste of phosphorus, chewed on match heads and started many destructive fires. Even more serious was the problem of phosphorus poisoning among match workers and, to a lesser extent, users. The deadly, necrosis-causing "white"

Commercial matches came in eye-catching wrappers.
Warshaw Collection, National Museum of American History,
Smithsonian Institution Photo 90-8371.

phosphorus remained in use in American factories until 1911, when the Diamond Match Company successfully substituted nontoxic "red" phosphorus in its operations.[10]

Spurred on by these problems as well as by simple curiosity, investigators continued to search for better ways to start fires. Nineteenth-century physicists succeeded in producing flames with electrical currents and even with solar radiation enhanced by "burning glasses" that were occasionally made from ice. Nevertheless, friction has remained the dominant method for generating fire-starting heat in America. The celebrated matchbook was invented in 1892 and began to achieve great popularity as soon as its suitability for advertising became widely recognized. Coming full circle, contemporary cigarette lighters, the precursor of which was introduced in 1909, employ a type of flint-and-steel mechanism to spark a highly flammable liquid fuel into flame.

Maintaining Fires

Fuel is all around us, but it has to be gathered before it can be used efficiently to do productive work. The first gathering and storing activities pursued by primitive people may have been in relation to collecting fuel for their fires.[11] Unquestionably, this was a central and time-consuming occupation for their descendants. The typical New England household of the late eighteenth century burned an estimated thirty to forty cords of wood per year.[12] That level of consumption would have required cutting down more than an acre of trees annually. Farmers often maintained their own wood lots for this purpose and chopped away as needed, but even town dwellers who bought their wood from dealers had their work cut out for them. Firewood had to be split, stacked, and eventually hauled to the house. In their domestic manual of 1869, Catharine Beecher and Harriet Beecher Stowe recommended further that firewood be carefully organized, with separate piles maintained for green wood, dry wood, oven wood, kindling, and the small amounts of charcoal that were needed for broiling and ironing.[13] The author of the *American Domestic Cyclopedia* (1890) claimed that most women would be content simply with an ample supply of dry wood: "Some men say, 'Oh! anything will get burned if we bring it in.' Any man who amounts to anything knows that woman's work is harassing enough without tormenting her with green, wet wood. Whenever I see a large woodpile or, better still, a shed filled with finely split, dry wood, I know that man appreciates his wife, and his position in his household."[14] Whatever the system, the woodbox had to be filled, and the job fell primarily to the men and boys of the family.

Despite depletion of the nation's forests, reliance on wood continued throughout the nineteenth century. It has been estimated that between 1630 and 1930 approximately 12.5 billion cords of wood were consumed by the wood-fuel trade.[15] Wood served as the chief fuel for the railroad for forty years, and as late as 1900 the wood-burning stove was a standard option for homeowners. Nevertheless, the introduction of coal, which began to be widely available for home use after about 1820, altered the work patterns for many Americans.

Naturally, it was less exhausting and time-consuming to buy a load of coal than to chop down a tree. Coal fires were hardly labor-free, however. Heavy buckets of fuel had to be carried from the coal bin to the hearth or stove. Additional preparations included sorting the pieces of coal by size and sifting them to eliminate contaminating dust (any number of fancy devices were sold for the purpose). Retrieval of cinders from the fire for reuse was yet another ongoing task. Because the mining of coal was a commercial rather than a family enterprise and because by the middle of the nineteenth century many men had jobs outside the home, much of this work fell on the shoulders of the women in the house.[16]

So, in large measure, did fire-building tasks, although it is clear from contemporary accounts that these duties were often shared. The lighting of an ordinary wood fire on an open hearth required an extraordinary number of manipulations. The standard technique for arranging wood in a fireplace called for a sturdy foundation of large logs, preferably green ones. Typically, a huge backlog was rolled into position at the rear of the fireplace, with a slightly smaller log placed on top. A third log, often called the "fore stick," was then rolled into position in front, at which point additional fuel, preferably dry, could be piled on as needed. It was generally agreed that walnut, maple, hickory, and oak burned well and that chestnut and hemlock should be avoided. Regarding the disposition of this fuel, however, there was anything but agreement, as revealed in Harriet Beecher Stowe's account of one family's approach to fire building in her novel *Oldtown Folks*:

> The rearing of the ample pile thereupon was a matter of no small architectural skill, and all the ruling members of our family circle had their own opinions about its erection, which they maintained with the zeal and pertinacity which become earnest people. My grand-father, with his grave smile, insisted that he was the only reasonable fire-builder of the establishment; but when he had arranged his sticks in the most methodical order, my grandmother would be sure to rush out with a thump here and a twitch there, and divers incoherent exclamations tending to imply that men never knew how to build a fire. Frequently her intense zeal for immediate effect would end in a general rout and roll of the sticks in all directions, with puffs of smoke down the chimney, requiring the setting open of the outside door; and then Aunt Lois would come to the rescue, and, with a face severe with determination, tear down the whole structure and rebuild from the foundation with exactest precision, but with an air that cast volumes of contempt on all that had gone before. The fact is, that there is no little nook of domestic life which gives snug harbor to so much self-will and self-righteousness as the family hearth; and this is particularly the case with wood fires, because, from the miscellaneous nature of the material, and the sprightly activity of the combustion, there is a constant occasion for tending and alteration, and so a vast field for individual opinion.[17]

As this discussion implies, fire tending occasionally evolved into a plea-surable form of recreation. Most of the time, however, those burning logs had to cook food, boil water, and supply the heat for domestic life. The shifting and poking of sticks, therefore, constituted basic and ongoing household work—work rendered all the more arduous by the need to lift heavy pots and bend and stoop over hot coals.

The cast-iron cookstove, which was perfected in 1815 and introduced to consumers soon thereafter, promised relief from some of these fire-related tasks. Boasting movable grates, adjustable dampers, and sturdy fireboxes, stoves offered the possibility of increased control over the process of combustion. Yet work was hardly eliminated. The fuel, whether wood or coal, had to be hauled to the stove, deposited, and lighted. Realignment of the dampers was a constant necessity, especially when weather conditions played havoc with the air currents. The accumulated ashes required raking and disposal. A study conducted by the School of Housekeeping in Boston in 1899 found that it took almost an hour a day to care for a modern coal-burning stove. In a six-day period the stoves chores broke down as follows: sifting ashes (20 minutes), laying fires (24 minutes), tending fires (1 hour and 48 minutes), emptying ashes (30 minutes), carrying coals (15 minutes), blacking the stove (2 hours and 9 minutes).[18] It is hardly surprising that one woman referred to the stove as "the black beast of her despair." "There it stands," wrote Caroline Corbin in 1873, "sullen, immovable, inexorable."[19] Perhaps the only consolation was that other fire duties may have seemed simple by comparison.

But other duties there were. Before the introduction of the self-consuming wick in the 1820s, candles required regular snuffing in order to burn properly. This did not mean "extinguishing," as is generally believed today, but rather "trimming" to keep the wick free of charred bits. According to one estimate, it had to be done as many as forty times an hour.[20] Oil lamps not only had to be filled and lit, but they worked best if their wicks were trimmed every day and their chimneys and shades washed at least once a week. Kerosene lamp chimneys benefited from daily cleansing.

Countless other tasks related to fires taxed the homeowner and industrialist alike. These included the storage and disposal of matches, the removal of ashes (these could sometimes be sold), and, last but not least, the perpetual cleaning of the sooty and ash-laden floors and work surfaces that necessarily developed anywhere near an open flame.

Periodic Maintenance

In addition to the regular chores, fire users had to provide for periodic maintenance of their equipment. Such requirements were most pressing in large-scale manufacturing operations where fires tended to be large and temperatures high. At the Saginaw saltworks in Michigan during the 1860s, for instance, huge fires were maintained night and day for five days straight in order to evaporate the brine most efficiently. Then on Saturday night the fires were permitted to "go down" so that Monday's work could be devoted to repairs on the furnace, kettles, and other tools. On Monday evening the fires were rekindled for another cycle.[21] A similar pattern was followed at many of the nation's charcoal blast furnaces, although some facilities attempted to maximize production by maintaining operations as

MODERN SIX-HOLE STEEL RANGE.

With Reservoir and High Closet.

FOR COAL OR WOOD

This Range possesses all the best features of a strictly first-class Up-to-Date Steel Range and is the best bargain for the money ever offered.   Prices do not include Pipe or Cooking Utensils.

ASBESTOS LINED
throughout to prevent the heat from entering into the room and gives oven full benefit of same.

DUPLEX GRATE
operates in oval fire box and can be removed, without disturbing the linings, through the front grate door.

NICKEL TRIMMED
throughout as shown in cut. Nickel is of a bright, lustrous finish.

MEASUREMENTS
we do not include the swell of the doors. Don't be deceived by the numbers marked on competitive stoves.

FLUE STRIPS
are cast iron to prevent them from rusting out. Also bottom of pipe flue is cast iron.

THE PRICE
is the only thing cheap about this range, for the very best materials are used in its construction, and range is perfectly constructed.

CAST IRON
reservoir casing will never rust out like steel, for cast iron is not affected by creosote.

LININGS
are made of the best selected pig iron, extra heavy, and will outlast any malleable top range, called "steel" to deceive.

Six-Hole Steel Range With Reservoir and High Closet.
If desired sent C. O. D. add $1.00 to the following cash prices.

Cat No.	Size.	Size of Lids.	Size of Oven inches.	Size of Top.	Height of Main Top	Length Fire Box.	Weight.	Price.
R M C 7-16	7-16	7	16x21x14	45x29	30	25	431	$26 00
R M C 8-16	8-16	8	16x21x14	45x29	30	25	432	26 05
R M C 8-18	8-18	8	18x21x14	47x29	30	25	443	28 05
R M C 9-18	9-18	9	18x21x14	47x29	30	25	444	28 10
R M C 8-20	8-20	8	20x21x14	49x29	30	25	459	30 55
R M C 9-20	9-20	9	20x21x14	49x29	30	25	460	30 60

With Steel Reservoir Casing $1.00 less.   Water Front $3.00 extra.

<u>WE GUARANTEE</u> any of our stoves to reach you in as good condition as when shipped from our foundry, but in event of breakage please notify us immediately and we will furnish repairs to replace broken castings promptly without charge.

<u>REPAIRS.</u> See directions for ordering repairs on the front page of our catalogue, and please be very careful to follow the directions given in every particular to avoid mistakes.

Major stove companies, such as the Modern Stove Manufacturing Company of Chicago, issued elaborately illustrated catalogs with dozens of designs. A typical page lists the many features of the latest models ca. 1900.

long as major problems could be averted. It was not unusual during the late eighteenth and early nineteenth centuries for furnaces to operate without interruption for seven to nine months, and a few enterprises, such as the Reading Furnace in Chester County, Pennsylvania, maintained steady production for more than twelve months at a time.[22]

There were no such luxuries in most American glassworks of the early period. The typical glass furnace was designed to accommodate from six to ten clay pots, each of which held as much as a ton or more of molten metal. Owing to the intense heat generated by the central fire, these containers lasted only about six weeks. The replacement process, which called for bringing the pots to white heat and then placing them over the blazing fire, was "one of the most arduous and dangerous duties in a glasshouse."[23]

Domestic fires were less intense but hardly less labor-intensive. Cast-iron firebacks and sturdy iron firedogs helped to minimize the damage to fireplace masonry, but repairs were an ongoing fact of life. So was hearth cleaning. In addition to the daily chore of sweeping, a thorough washing was necessary at intervals to keep the fire area tidy. For large, "dirty" families, Lydia Maria Child recommended nothing more than soap and water with a bit of lamp oil rubbed in. But, "if you wish to preserve the beauty of a freestone hearth," she wrote in 1833, "buy a quantity of free-stone powder of the stone-cutter, and rub on a portion of it wet, after you have washed your hearth in hot water. When it is dry, brush it off, and it will look like new stone." Child explained that brick work was best kept clean with "redding," water, and a bit of skim milk, whereas the more elegant marble fireplaces benefited from regular dusting, oiling, and drying with a soft cloth. Fireplace tools also required a judicious pairing of composition with cleanser on a frequent schedule.[24]

The inner walls of the chimney demanded attention as well. Frequent removal of the soot and unburned fuel that inevitably accumulated there was a prudent fire prevention measure and was required by law in many localities. There were a number a methods by which householders could do the cleaning themselves. Among the most popular were brushing the chimney with a long pole, dropping a chain down the shaft and allowing it to jar the debris loose, shoving a goose up the chimney so that its flapping wings would accomplish the same thing, and, finally, deliberately setting the entire chimney on fire. (The latter approach, though popular in Europe, was extremely dangerous in wooden American cities and was severely restricted by local legislation.)

By any method, though, chimney cleaning was a dirty business, and the services of professional sweeps became an attractive alternative to self-cleaning. The public's need for satisfactory levels of performance often led communities to regulate rates and standards. Rarely, however, did such legislation extend to the protection of young sweeps—typically black boys apprenticed as young as eight years of age—who scraped and brushed their way from the hearth to the roof. Throughout the nineteenth century, inventors constructed and patented a number of mechanical devices that promised to liberate these "climbing boys" from what George Appleton termed a "dreadful trade," but these options were rarely pursued. Inadequate child labor laws and outright racism meant that human beings, rather than machines, kept the nation's chimneys clean until long after the advent of central heating.[25]

By that time, most American families were struggling with periodic maintenance problems of another sort, namely, those imposed by cast-iron stoves. As household manuals firmly pointed out, periodic "blacking" of the stove's surface was essential to prevent rusting. In addition, cooks were advised to remove ashes from the flues behind the ovens at least once a month and to clean drafts and the chimney four times a year. In the case of heating stoves that were in use only part of the year, summer afforded a perfect opportunity for repair work. Denison Olmsted, a Yale professor who had a great interest in promoting the use of anthracite in the early 1830s, described some tasks that he believed should be performed on anthracite stoves, but the suggestions were appropriate for other fuel burners as well:

> The stove and pipe should be taken down and brushed clean, and if a white wash of lime with a mixture of fine white sand be applied, it will contribute much to the durability of the apparatus. When thus taken care of, sheet-iron stoves may be made to last many years. There is, moreover, another important reason for clearing a stove-pipe of all deposits or incrustations formed on its interior surface. These not only impair the draught, but, being non-conductors, they greatly diminish the power of the metal to absorb and diffuse heat.

> At the close of the season for fires, anthracite stoves frequently require more or less repairing. The lining, perhaps, is broken, or concretions are formed on it that cannot easily be detached, or some of the dampers are out of order. It will generally be found advantageous to commit the whole work of cleaning and refitting to the stove-dealer; but, as this may not always be convenient, the following hints may be useful. Where the lining is broken, as it frequently is in endeavoring to separate the slag or concretions, it may be repaired by a lute made as follows. Procure (of the stove-dealers) a piece of *fire* clay; pulverize it in a mortar, and add twice its weight of clean white sand; add a little water, and beat the mass until it has the consistence of putty. Brush off the dust from the lining, and wet the part to which

William H. Rau, Publisher.

Philadelphia, Pa.

17213　　When a Man's Married His Troubles Begin.

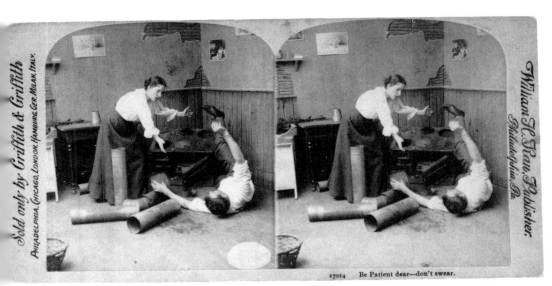

William H. Rau, Publisher.

Philadelphia, Pa.

17214　　Be Patient dear—don't swear.

Stove maintenance was a time-consuming—and at times frustrating—part of domestic life.
These humorous stereographs, titled "When a Man's Married His Troubles Begin" and
"Be Patient dear—don't swear," were part of a series published by
Griffith and Griffith, Philadelphia, ca. 1895.

the lute is to be applied. Finally, press on the lute firmly, smooth it down with a broad knife, and suffer it to get dry before the fire is kindled.[26]

Seasons of Fire

Olmsted's reference to the "close of the season for fires" implies that fire users enjoyed periodic reprieves from their fire maintenance chores. Up to a point, that notion has merit. Unquestionably, the end of winter meant the cooling—and even dismantling—of large numbers of domestic heating stoves, as well as the temporary disappearance of those that served trains, shops, and other public places. Fewer stoves meant fewer fires—until, of course, the cycle began again.

Other, less obvious, ways to eliminate fire chores were also available. One common response was the adoption of multipurpose contrivances that required only one fire. In a sense, this approach was a natural out-

The Edwards Parlor Lamp Stove.

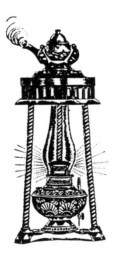

AS A COOKER.

This cut shows the stove in use as a cooking stove, for which purpose it has received the highest recommendation from ladies who have used it for light house keeping.

THE PARLOR LAMP STOVE

. . . . *Is "a thing of beauty and a joy forever."*

There is no need of keeping yourself poor, paying exorbitant coal and gas bills, for this Lamp Stove is a perfect substitute for both, as it gives a 300 candle-power light, and heat enough for any ordinary parlor, sitting room or office.

It is something entirely new, and needs only to be seen to be appreciated and desired by all.

There is no oil burning device on the market that is such a POWERFUL HEATER *and nothing to compare with it in* SPLENDOR.

EDWARDS PARLOR LAMP STOVE CO.
9 & 11 NORTH CANAL STREET,
CHICAGO, ILL.

The Edwards Parlor Lamp Stove Company offered the public a number of innovative devices, including this handsome lamp that doubled as a stove.

growth of the open-hearth school of fire building, whereby a single fire provided light, warmth, and heat for cooking. The concept evolved over the decades and finally came into its own during the second half of the nineteenth century, when double-duty devices seemed to flourish in American society generally: sofa beds, music stands that transformed into ornamental swords, and any number of brass band instruments that could be switched from one key another with the flick of a knob.

The dual-purpose fire-powered devices created by American manufacturers were almost as intriguing as these novelties. Although kitchen cook stoves were by definition heating stoves, many distributors also promoted parlor stoves that doubled as cooking devices. "Put supper on in sitting room and did not build a fire in kitchen," noted one woman who enthusiastically took advantage of such a device many times during the winter of 1867.[27] Even more common were lamps that doubled as food warmers. It was simplest to achieve this feat on a small scale in the manner of Clarke's highly touted "Pyramid Food Warmer." "When nights are dark, then think of Clarke," urged the advertising rhyme that highlighted this combination night-light and milk warmer.[28] Larger lamps could process food more effectively, and to this end manufacturers offered a variety of products and adaptive parts. Turn-of-the-century consumers, for instance, could acquire a "lamp chimney stove" from Sears for just three cents. The device fit any ordinary crimped-top lamp chimney and purportedly boiled water in just a few minutes.[29]

An oriental theme (complete with smoking Chinaman) surrounds this ca. 1880 trade card for the coal-fired Astral stove. Like many parlor stoves of the day, this model heated water while it warmed the room.

The idea of using kitchen fires to create a perpetual supply of hot water was so sensible that hot-water tanks became common accessories for cook stoves during the last third of the nineteenth century. Less ordinary were parlor lamps that doubled as furnaces, but such items were not unappreciated. "The Falls Heater with an ordinary library lamp kept my library, 15 × 14 ft., up to 70 degrees with no trouble," wrote one satisfied purchaser. Taking advantage of the fact that oil lamps gave off heat as well as light, the enterprising creators of the Falls Heater simply designed a reflecting apparatus that, when placed over a lamp chimney, forced the heat out into the room. Their advertising campaign focused on the advantages of the product at once: "Saves money, saves time, saves coal." Most important of all, "Saves Making a Fire!"[30]

These many products appealed to the public because they embodied the "can-do" approach to problem solving that was so highly valued in America during the nineteenth century. There was, however, yet another way to cut down on fire-tending chores: to do without. Ever since colonial days, homemakers had saved both fuel and work by closing off various rooms and concentrating their energies on a single, centrally located fire.[31] Such a philosophy worked as well on a daily basis as it did seasonally. When it was too much trouble to stir up a flame, many people simply did not bother. That was Rachel Haskell's choice one chilly March evening. "Ella and I washed dishes in [the] cold kitchen after every one left," she wrote after an evening of entertaining.[32] Peggy Rawle tried the same approach with regard to lighting devices. The moon was so bright one spring night in 1781 that she sat quite happily in the parlor "without a candle."[33] Other families adopted similar patterns according to their needs—it was not uncommon for people to prepare for bed in the dark—and occasionally they even had firmly entrenched social customs to back them up. It seems quite likely, for instance, that the restrictions against cooking on the Sabbath that mill-worker Lucy Larcom and other New Englanders remember so fondly from childhood were observed not only because of religious strictures but also because of the appeal of an enforced cessation of fireside work.

But these were temporary measures at best. In the broadest sense, the "season for fires" was eternal. Fire building was required for cooking, lighting, industry, and transportation all year long. Likewise, as everyone knew, the work associated with these fires continued with relentless intensity.

FIRE TENDING: THE INSULATING FACTOR

Although the need for fire remained unchanged, America's fire habits changed dramatically over the course of the nineteenth century. Technological innovations, the introduction of new fuels, and the transition to a

Technological developments of the nineteenth century tended to insulate the user from flame. This logical progression is illustrated in trade cards by the Florence Machine Company, ca. 1880.

consumer-based society conspired to alter forever the ways people handled and thought about the open flame. There are many ways to chronicle these momentous changes, but no approach serves as well as to focus on the process of "insulation"—how domestic fire users in particular became increasingly insulated from the open flame and the mechanics of fire manipulation.

Technological Advances

TOOLS

Prior to the invention of the chimney in the eleventh century, indoor fires were built in the center of a room and vented through a smoke hole in the roof. Essentially, fire users lived inside their fireplaces. Owing to its simplicity, this ancient design was reproduced by European settlers in America whenever swift construction or temporary structures were required. Later, when the time was taken to build proper hearths and chimneys (or even "improper" ones of sticks and mud), the central fireplace tended to dominate most homes and continued to do so well into the nineteenth century.[34] "The house seems to be built against the chimney, if one may judge by its size and prominence as the most important feature," wrote an Englishman who was struck by this phenomenon while visiting Kansas in the 1850s.[35] Nathaniel Hawthorne describes the huge central fireplace in *The*

House of the Seven Gables in much the same way—as a living, breathing, and altogether unavoidable presence.

Both depictions are historically accurate. The utter centrality of the fireplace to a family's life and work cannot be overstated. Consider the situation circa 1800. In New England the fireplace was constructed in the center of the house, where it could provide maximum heating effectiveness; but even in the South, where fireplaces tended to be placed on external walls, their gaping presence was apparent. People referred to their "fire rooms," where they baked "fire cakes" and told "fireside stories." They picked up pieces of fire and moved it around by hand to light their candles, rushlights, and grease lamps and to fill their warming pans and irons. They also picked up their chairs and chores and moved them toward the beckoning flames of the open hearth. During the winter months, in fact, people were virtually tethered to their firesides. They might stray to do some work or to sleep, but the heat and light would always reclaim them. And when it did, the raw energy of the open flame continually assaulted their senses, irritating their eyes with the flickering light and teasing their metabolisms with notoriously uneven heat.

It was in the realm of open-hearth cooking, however, that the proximity of the user to the flame was most apparent. Hearths varied in size (it was not unusual for them to measure six feet wide by three feet deep); but whatever the dimensions, the cook had to stand *in* the hearth to work. And work it was. There was not just one fire to cook on but several, a situation that required constant maneuvering, adjusting, and vigilance. Blazing logs had to be kept alive for roasting meat, whereas hot coals had to be shoveled on top of Dutch ovens for baking bread. Baking in an oven was a separate process altogether. The cook had to light a new fire (preferably using light-weight "ovenwood"), allow it to heat up, and then rake the fuel out before the food could be introduced. Gauging the right temperature was a matter of experience, but virtually all of the early tests called for the cook to be physically involved with the fire—to feel the heat with her hand or to watch how fast flour scorched when strewn on the bottom of the oven. That this relationship was potentially dangerous is suggested by some of the early receipt books, which refer candidly to "furious" and "violent" heat.[36] Nevertheless, despite the hazards, open-hearth cooking was very much a family affair, and even children were permitted to turn spits and toast bread. "If I could only be allowed to blow the bellows . . . when the fire began to get low," recalled Lucy Larcom of her childhood, "I was a happy girl."[37]

As it turned out, Larcom did not remain a happy girl for long. Over the course of her lifetime, the open fires that had given her so much pleasure during her youth were gradually replaced by new kinds of fires that were at once more efficient and less visible. In fact, the open hearth had three lines

This scene was described as "primitive cooking" when Benjamin Butterworth produced his survey, *The Growth of Industrial Art*, in 1892. Many hearths were larger and lacked the protective grating.

of descent, each leading to improved methods of heating, cooking, or lighting and all entailing increased insulation of the user from the flame.

Consider what happened to domestic heating fires. Some of the early improvements introduced only minor changes in people's fire habits. Iron fireplaces, of which Benjamin Franklin's famous "Pennsylvanian fireplace" of 1742 was one, fit into existing fireplaces and maintained a view of the open flame while simultaneously conserving fuel. Similarly, coal-based hearth heating arrangements, though capable of achieving a steadier and more intense heat than an open wood fire, required merely the addition of specially designed grates to restrain and ventilate the fuel, which was still fully visible.

The appearance of the cast-iron stove, however, was the revolutionary development that moved heating fires farther and farther away from the user—initially, to enclosed metal boxes in various rooms and finally to a single basement fire called a furnace. Such devices had been used in America as early as the 1660s, but it was not until the 1830s that the innovations of Eliphalet Nott and others began to make them attractive to homeowners[38] and not until the 1850s that manufacturers could offer a variety of models at reasonable prices. Their impact on fire habits was uneven at first. Although many of these stoves enclosed the fire from view—one of Nott's claims to fame was that his invention was the first anthracite-burning stove to enclose the fire fully and satisfactorily—fire chores inevitably per-

By 1897, when this photograph was taken, fire was hidden away within a stove, but it continued to dominate the room.

Stoves were prominently displayed in stores carrying a variety of general household items. The location of C. U. Grenwalt's Hardware Store is not given, but the view is typical of any turn-of-the-century American town.

The Cortland Howe Ventilator, introduced in the 1890s, was one
of many central heating schemes devised for American homes.
This domestic scene shows the mother and child at a great
distance from their fire.

sisted. Even with gravity-fed systems, the hoppers had to be filled, the flues
adjusted, and the ashes removed by hand. Gradually, however, hands-on
manipulation of home heating systems receded. Central heating, which
was installed in more expensive homes as early as the 1860s and which
spread more generally during the 1890s, eliminated all fires but one. Automatic stokers (available in the early 1900s) and electromagnetic thermo-
stats (introduced in the 1920s) finally freed the homeowner from having
to build fires to heat the home.

Though the details differ, the same progression can be chronicled for
cooking technologies. Cast-iron cook stoves, which had replaced the
open hearth in most homes by the time of the Civil War, provided a

physical barrier between the cook and the fire.[39] The user saw the flames only when necessary and the inside of the chimney rarely at all. The height of these devices also tended to exclude children from fire-tending chores, with the result that fire manipulation was decreasingly a focus of childhood education.

The evolution of lighting devices is more complex, owing to the wide variety of fuels and lamp designs, but a similar pattern is discernible. The open candle gave way to ever more enclosed sources of light. Glass chimneys, invented by the Swiss chemist Ami Argand in the late eighteenth century, protected the user from the flame even as they improved the efficiency of combustion.[40] Gas lighting technologies achieved a similar end, not only by positioning the gas jets high above the user but also through the use of ornate shades. The invention in 1885 of the Welsbach mantle, a treated cotton mesh cylinder that surrounded the flame and glowed with a brilliant incandescence, added yet another barrier.

FUELS

Even as users were increasingly insulated from the open flame by mechanical devices, they were also increasingly spared the hands-on acquisition of fuels. The progression from wood to coal to gas and petroleum-based fuels underscores this point.

Wood fires, whether on the open hearth or in a stove, demanded hours of preparatory work, from felling the timber to hauling it, chopping it, stacking it, and ultimately positioning it on the fire. It was basically a home-based operation in which each family member could take part. In many cases, the head of the household went out and chopped the trees personally. Even if wood was purchased, the subsequent cutting and carrying was the responsibility of the users. Adult males split the wood, children (particularly boys) gathered kindling and kept the woodbox full, and housewives transferred the wood to the fire as needed and coaxed it into flame. As William Chauncy Langdon has noted, wood was a fuel thoroughly integrated into family life: it came in through the door—much like a person.[41] Furthermore, the ubiquitous woodshed acquired social connotations that went far beyond its original purpose.[42]

This cozy, self-contained arrangement was duplicated with a number of other fuels—most notably the prairie grasses and buffalo dung that fed the fires of many Midwesterners of the mid-nineteenth century.[43] Yet, family fuel-gathering rituals collapsed almost completely with the advent of coal. Coal, both domestic and imported, was used sporadically in America during the eighteenth century, but not until the opening of the anthracite coalfields in Pennsylvania in the 1820s was the fuel marketed aggressively. Thereafter, it increasingly took over home heating and cooking, and by

WINTER IN THE COUNTRY.

Wood provided a ready energy source for many Americans during the nineteenth century. Gathering and cutting firewood were time-consuming tasks. Peters Collection, National Museum of American History, Smithsonian Institution Photo 60421-C.

1880 almost all of the nation's railroads had switched. Like wood, this fuel required considerable manipulation by its users. It had to be cleaned and carried from the coal bin to the fireplace or stove. Once in place on its grates, it had to be monitored periodically to maintain effective burning, and its ashes had to be removed when the burning process was complete. Unlike wood, however, coal also required the labors of specialists: the miners who dug it out of the ground and the merchants who distributed it. Although the users had the last word, so to speak, they were directly involved in only the last stages of this fuel's very complicated transit from place of origin to hearth.

By their nonsolid nature, carbon-based fluids such as gas and oil precluded hands-on manipulation in all but the filling and lighting operations. From whale oil, coal gas, and "camphene" (a mixture of alcohol and redistilled turpentine), which appeared during the antebellum period, to the wide range of petroleum-based fuels that came to predominate in later decades, these fuels required extensive preparatory measures before they

The successful exploitation of petroleum in Pennsylvania in 1859 led to a boom business that provided lighting fuel for many Americans. This view shows a Pennsylvania well, ca. 1870. Stereograph by C. W. Woodward, Rochester, New York.

could be transformed into energy by the user. Much labor was involved in drilling wells, refining and purifying raw materials, constructing pipelines, and managing the complex operations. The user's role was forever transformed from energy producer to energy consumer.

Different people reached this state at different times, depending on their energy needs. Even so, the trend was set. Whereas most people produced most of their own energy when the nineteenth century began, this was no longer true as the century drew to a close. The age of the axe was irrevocably transformed into the age of the fuel bill.

Not surprisingly, fuel distributors described the advantages of their products in glowing terms. One natural gas company listed nine compelling reasons to switch to that fuel in the 1890s. Gas "saved labor" and was "always ready," they claimed. "A fire can be ready at a moment's notice," the advertising brochure added triumphantly.[44] That should have been good news to householders inured to fires that took many minutes to prepare. Nevertheless, the suspicion with which many people greeted the installation of gas meters in their homes (to "lie like a gas meter" was a popular expression) gives some indication of the alienation caused by these developments.

OTHER INSULATING FACTORS

New devices and new fuels were, in a sense, externally imposed modifiers of domestic and industrial fire habits. They required large-scale manufacturing and distribution networks for adequate implementation. In addition to these commercial efforts, however, individuals frequently devised their own ways of insulating themselves from the work, dangers, and perpetual annoyances of fire maintenance. The most common approach—and also the most effective one—involved hiring, coaxing, or otherwise compelling someone else to do the work. "In the morning the boy gets up and makes a fire by seven o'clock when I get up and make the coffee," wrote a San Francisco hotel owner, describing a common solution to the perennial problem of fire starting.[45] "There were public sawyers who did most of the wood-sawing" and offered their services "almost as soon as the wood was unloaded before your door," recalled William Dean Howells.[46] Adapting this idea to the academic world, chemistry professors at the University of Virginia routinely delegated the job of maintaining the laboratory fires— including the onerous task of working the outdoor bellows—to the student with the worst grades.

Many of these fire workers had specific occupational titles. There were woodmen to furnish or chop a family's supply of fuel, ashmen to collect ashes, and furnacemen to care for the basement heaters in large buildings and, to a lesser extent, homes. The term *chauffeur*, which derives from the French word *chauffer*, meaning to heat, was occasionally applied to the stokers of fires on steamboats and locomotives, after which it became the logical designation for the mechanic who stoked the steam engines in early automobiles.[47] More commonly, however, stokers of steam engines and industrial fires were called firemen, in direct reference to their responsibility of keeping the fires burning. On the home front, by contrast, fire workers tended to be called, simply, servants.

It is impossible to know how many households had domestic help during the last century. It has been suggested that for the period 1660–1860 most adults had some kind of assistance with their domestic work at some time during their lives and that later, as the population increased, as many as 15 to 30 percent of urban American homes employed at least one live-in servant.[48] What is certain is that many of these employees routinely performed fire duties.

Philip Fithian's diary gives a curious glimpse of this reality. Early on Christmas morning in 1773, his "boy" Nelson showed up as usual to light his fire. Because it was Christmas, Nelson also received a gift. Moments later, the pattern was repeated. "The Fellow who makes the Fire in our

School Room . . . entered my chamber," wrote Fithian. He too wanted a reward for his services.[49]

And service it was. "Light the fire; clean the fender, fire irons and hearth; take up the ashes, sweep the carpet, shake the hearth-rug and lay it down again." These were some of the household chores delineated by Sarah Josepha Hale in her *New Household Receipt-Book* (1853) and typically assigned to domestics at mid-century.[50] The pattern spread during the remainder of the century. As men increasingly found employment away from the home and were therefore unavailable to help with domestic fire maintenance, and as cooking and heating fuel shifted to coal, which was heavier and dirtier and generally less "genteel" than wood, fire work was increasingly assigned to the hired help. Contrary to the advice of some household experts, even the lighter chores, such as trimming lamps and washing lamp chimneys, were frequently handed over to servants by housewives who did not want to be bothered with dirty, menial labor. *Bother* is a key word here. It is likely that one reason for the phenomenal popularity of boardinghouses during the nineteenth century (it has been estimated that more than 70 percent of the population lived in boardinghouses at some time of their lives) was that in such an environment residents not only evaded most fire chores altogether but also avoided having to oversee the servants who did the work in their stead.[51]

There is no doubt, of course, that many jobs required enormous skill in fire manipulation. Founders at ironworks, gaffers at glassworks, boilers at saltworks, and blacksmiths at their forges knew from experience how to "read" a fire and use it for their purposes. Housewives who relinquished the basic cooking duties to their servants frequently retained the more exacting baking chores for themselves. On the other hand, it is clear that many fire jobs rested firmly in the category of "common labor." Firemen at mid-century saltworks were paid at the same rate as ordinary hands. Firemen on locomotives, though they worked with the engineer, were uniformly accorded lower status and pay. And in schoolhouses and factories across the country, boys started the morning fires. American civilization depended on fire, but the fires themselves clearly depended on the exertions of the common man, woman, and child.

Many members of the middle and upper classes found solace—or at least a saving of labor—in this situation. There was another way for fire users to remove themselves from the flame, however. This somewhat oblique approach to the problem might be called "insulation through ornamentation" because it relied not so much on distance as on disguise. Whereas servants and advances in technology served to shield users from the flame physically, patterns of decoration created a kind of barrier of the mind.

Although few practitioners of the art went as far as those misguided people who decorated coal scuttles with flimsy paper photographs,[52] the "tasteful" ornamentation of fire tools found nearly universal support throughout the nineteenth century as an effective means of de-emphasizing the utilitarian aspects of domestic energy-producing equipment through the imposition of a veneer of artistry.

It has been claimed that household decoration itself began with the fireplace.[53] Certainly, by the time of the Federal period, fireplaces and the associated mantelpieces and chimneys had undergone extensive decorative embellishment in upper-class homes. By the middle of the nineteenth century the movement had spread to more modest dwellings, where elaborate shelving arrangements, decorative tiling, and carved wooden borders helped to beautify the fireplace area. As advice books were quick to point out, the process need not be expensive. The point was simply to provide the fire with a "fitting habitation."[54] Augmenting the enhancement of the hearth itself were legions of decorated implements and tools. These included hand-painted bellows and fire boards (the latter covered the opening of the fireplace during summer), cast-iron firebacks depicting animals and stories, andirons in a variety of shapes (Continental soldiers were especially popular at one point), and even elegant brass pokers and shovels that were designed for show rather than use.

Lighting devices received similar treatment over the years. From the early betty lamps, lovingly enhanced with whimsical animals, hearts, and curlicues, to the more sophisticated oil and kerosene lamps with ornate bases and handsome glass chimneys, illuminating mechanisms consistently transcended their utilitarian origins as vessels of combustion to become works of art. Although cast-iron stoves at first presented a challenge to proponents of this trend, aesthetics quickly tamed these hulking black beasts as well. Marketed with such names as the "Jewel" and the "Art Garland," parlor stoves were embellished throughout the second half of the nineteenth century with scrollwork, metallic orbs, and decorative urns. Despite the protests of some housewives, even cookstoves were subjected to ornamentation in the form of textured iron doors and fluted legs. "Come! & look it over," beckoned a typical stove advertisement of the day. "You will be convinced, the day has passed when a stove has to be a black unsightly object, only to be tolerated as a necessity. That has all changed, & the most attractive piece of furniture in a room is a Glenwood Stove."[55]

This philosophy served as the foundation of the decorative movement as a whole. If a fire-making device looked less like a tool and more like a nicety, then the fire maker would undoubtedly feel less like an overworked producer of energy and more like an enlightened and civilized user. A

Every nineteenth-century American home featured fire technology, although this fancy parlor stove hides the utilitarian purpose of the device fairly well. Note the decorative draping and knick-knacks that conceal the old mantelpiece behind the stove.

person could be burned by a pretty stove just as easily as by an ugly one, but an overlay of decoration transformed the overpowering rawness of a fire into genteel "warmth" and "radiant beauty."

Radiator manufacturers of the early twentieth century embraced this approach enthusiastically. The National Radiator Company's Aero models were "artistically designed, with slender, graceful lines and beautiful proportions." As the company's 1926 catalog put it, "Here is a warming device harmonizing perfectly with the furnishings of the daintiest room, and at the same time possessing an inherent refinement of design that enables it to bear the most critical scrutiny."[56] By this time, of course, the insulation of users from the flame was almost complete. The fire was hidden away in the furnace in the basement and users' fire duties were vastly reduced.

Over the course of the nineteenth century, manufacturing was gradually wrested from the hands of the individual. Textiles and food, clothing and toys, medicine and furnishings—all were produced primarily at home during the eighteenth century and all were purchased from commercial establishments on a large scale by the end of the nineteenth. Energy is, in a sense, just another of these commodities. And, as in other areas of daily life, the changes in its mode of manufacture and distribution had profound consequences for American society at large.

By 1920, when this advertisement by the National Radiator Company appeared, some families had the option of gathering around the radiator instead of the stove.

The Making and Care of a Fire

Remove the covers, and brush the ashes from inside the top of the stove into the fire-box. Replace the covers, close the dampers, and turn over the grate. Shake the lower grate, letting the ashes sift through into the ash-pan. When the dust ceases to rise, brush out the oven, remove the cinders from the lower grate, and reserve them to burn again. When taken out in this way, the ashes in the pan will not require sifting. If there be no lower grate, remove the ashes and cinders together, and sift them. Pick over the cinders carefully, and throw out any stones, slaty pieces, or bits of clinker. These should never be burned, as they injure the lining of the fire-box; but any pieces of half-burned coal should be saved. Always take out the ashes before lighting the fire, for if they are left in the pan, sparks and lighted coals will drop into them. It is then highly imprudent to remove them, unless they are to be placed in a fire-proof ash receiver. Fires have often been occasioned by careless storing of hot ashes.

Put into the fire-box, first, shavings or loose rolls of newspaper, letting them come close to the front; then fine pine kindlings, arranged crosswise, that the air may circulate freely between the pieces; be careful to have them touch each end of the fire-box that the coal may not drop through to the grate. Then put on enough hard wood, arranged in the same manner, to come to the top of the fire-box. Put on the covers, open the dampers, and brush the dust off the stove.

Moisten some stove-polish with cold water, and put it on the stove with the "dauber." Rub the blacking in thoroughly, then light the paper from below the grate, and while the fire is kindling polish the stove with the dry polishing brush. Blacken the stove while it is cold, but polish as it begins to heat.

When the wood is well kindled, put in a few more pieces of hard wood, and press the coals down to the grate. Put on coal enough to cover the wood, and when this has kindled fill the fire-box to the top of the lining. By making sure that the hard wood kindles first, and adding the coal gradually, much trouble is saved; for unless the kindling be well seasoned, and part of it hard wood, and plenty of it used, it will either not kindle or it will burn out before the hard coal kindles, and then the coal must be removed and the fire rebuilt. The blazing heat from the wood alone warms the stove, and the oven quickly becomes hot. If you have charcoal or Franklin coal, it may be put on at first with the wood.

When the blue flame is no longer seen, close the oven damper; and as soon as the coal is burning freely, shut the front damper. Then regulate the fire by the slide or damper in the pipe.

While making and watching the fire, empty the teakettle, wipe out the inside, fill it and the reservoir with fresh water,—never from the hot-water tank,—finish polishing the sides and back of the range, and brush up the hearth and floor.

When a hot fire is needed for several hours, add a sprinkling of new coal before the first has burned out, and add to it often enough to keep the fire at a uniform heat. Be careful not to add enough to cover and thus check the fire, and never have the coal above the top of the lining.

When the fire is not needed for the present, add a little fresh coal, and close all the dampers in two or three minutes, or as soon as the blue flame disappears. Never shut off all the draught on a red-hot fire without putting on a little fresh coal, if you wish to keep it in good condition to use again. It is important to remember that when all the coals are red they are nearly burned out, and will not give out heat for so long a time as when partly black and partly red.

To quicken an old fire, open all the dampers; and if the coal is black or only partly burned on top, pick out the ashes underneath with the poker, and when it begins to burn more freely add a sprinkling of coal and shake the grate. Keep the grate free from ashes when a very hot oven is needed. But if the old fire has burned so low that all the coals look red or ashy, always put a few pieces of small coal on the red coals, and when these are burning add a few more carefully; then shake the grate gently, or pick out the ashes. If you shake a whity-red or dying fire, the ashes fly up and settle on the coals and put out the little life there is in them.

During cold weather, or when a fire is required for heating purposes as well as for cooking, it is more economical, with most first-class stoves, to keep the fire night and day, letting it go out occasionally if the grate become clogged. But when it is no longer wanted for either purpose, turn the grate over at once that there may be no unnecessary burning of the coal.[19]

(Mrs. D. A. [Mary Johnson] Lincoln,
*Boston School Kitchen Text-Book: Lessons in Cooking
for the Use of Classes in Public and Industrial Schools*
[Boston: Roberts Brothers, 1888], pp. 7–9)

6

Understanding Fire

The world had fire long before it had scientists, and scientists puzzled over fire long before they understood the mysteries of heat and flame. Alchemists and their successors contributed both information and a sense of urgency to these inquiries. Perpetuating the tradition of the ancient Hermetic treatises, these investigators relied on the transforming power of fire to help them take things apart. The more they learned about fire, the more they stood to learn about how the world was put together.

Most citizens had no such concerns—they had no need. For them fire was warmth and light, and they could enjoy these benefits without reference to chemical processes or natural laws. Here, then, was yet another way in which fire users were insulated from their energy source. This barrier of ignorance has numerous modern parallels—the driver who knows nothing about Newton's Laws, the microwave cook who knows nothing about electromagnetic radiation. Yet, for fire users of the colonial period there was an added level of complexity because *no one* understood fire very well—not the college professors, not the fireplace builders, not the firemen.

There were plenty of theories, of course. Some of the more popular explanations—the notion that fire was an element, for example—had been around since antiquity. But there were obvious deficiencies in the old concepts and a corresponding lack of coherence with the new. Around the time of the American Revolution, however, investigators in both Europe and America focused increasingly on the puzzle that was fire and began to fashion what have become our modern views on the process of combustion. The work was accomplished haltingly and without any unified attempt to enlighten the masses. Nevertheless, as the specialists learned more about fire, their knowledge inevitably trickled into the public consciousness. Slowly, inexorably, society's perception of energy was altered. In the process, some fire users actually became *less* insulated from the flame during the nineteenth century because, at last, they had some hope of understanding it.

This chapter explores the subject of fire and science from three perspectives. The first section traces the evolution of the science of combustion; the second describes the related, but essentially separate, campaign for understanding and improving fuels. The third section provides an over-view of some of the ways in which this vast body of esoteric knowledge spilled over into the public consciousness.

THE SCIENCE OF FIRE

Fire researchers were driven by both natural curiosity and pragmatic concerns. Natural philosophers wanted to understand how fire fit into the cosmos of matter and energy, forces and motion. Their experiments were designed to discover the laws governing heat and flame. Meanwhile, engineers used the hard-won empirical knowledge of factory and forge in their search for the most efficient means to employ heat and to exploit valuable fuels. In these separate efforts to deduce the nature of fire, both philosophical and practical researchers focused on two intertwined questions: What is heat? What is flame?

The Nature of Heat

Isaac Newton's *Philosophiae naturalis principia mathematica* (1687) provided a clear and compelling model for discovering the workings of the physical universe. Newton's three laws of motion and his concept of universal gravitation systematized the study of matter and forces while they unified observations of celestial and terrestrial phenomena. Orbiting planets and falling apples, events that appeared distinct and unrelated to his contemporaries, were seen by Newton as two manifestations of the same great underlying principles. Newton's vision transformed science by suggesting the existence of a few universal laws that apply to all systems. Following Newton, the ultimate objective of science was to search for these overarching principles.[1]

Newton provided a model for researchers hoping to characterize forces through the careful observation of phenomena. Electricity and magnetism, for example, were codified with this systematic approach. Yet the Newtonian framework was ill-suited for deducing the nature of heat and light, two of the so-called imponderable fluids. Neither heat nor light seemed to have mass, and neither appeared capable of exerting a force. A different set of rules was required.

THE PROPERTIES OF HEAT

Investigators started their quest for understanding by describing the behavior of heat. A key property of heat seemed to be its ability to flow from one place to another by any of three mechanisms: *conduction*, the shift of heat within a solid body, as when the handle of a metal spoon becomes hot soon after the opposite end is placed in a bowl of soup; *convection*, the movement of hot fluid masses, such as water brought to a rolling boil or a warm breeze in summer; and *radiation*, the process that transfers heat from the sun to the earth. Any theory of heat had to account for these three physically distinct forms of heat transfer.

Other properties of heat were obvious to early investigators as well. For instance, heat always seemed to flow from hot objects to cold ones, never the reverse. Fuels, when ignited, were known to release heat and lose mass. And most objects seemed to expand when heated and to contract when cooled. These properties suggested to most observers that heat was a substance that could be transferred from one body to another.

Joseph Black, an eighteenth-century Scottish chemist, quantified much of this everyday knowledge in his widely accepted version of the "caloric theory," which postulated the existence of an all-pervading fluid called caloric. Black believed that particles of caloric fluid are attracted to matter but repel each other. Caloric must have weight, he argued, because heated metals can gain weight.[2] Furthermore, the total amount of caloric in a closed system (one in which heat is neither added nor taken away) does not change. Caloric fluid is most concentrated in the best fuels, such as wood, natural gas, and coal.

The theory of heat as a fluid that passes from one body to another seems logical, for heat does behave like water in many ways. Heat flows from one place to another, for example; and like water, caloric comes in both pure and mixed forms. Heat can be stored in an inert form (as in wood) ready to be released, just as water is stored in everyday objects from rocks to a loaf of bread. Many objects expand when they soak up heat, just as a board will expand when it soaks up water.

The downfall of the caloric theory was its strict reliance on the principle of conservation. According to the theory, caloric—like water—could be transferred from one place to another, but the total number of fluid particles should remain unchanged. This requirement did not agree with the practical experience of American-born Sir Benjamin Thompson (Count Rumford), whose rich and varied life included a term as superintendent of cannon boring at Munich's military arsenal late in the eighteenth century.[3] Thompson noted that brass became hot while it was being machined, but he found that the amount of frictional heat generated depended on the sharpness of the cutting tools, not on the quantity of brass. Dull tools created a great deal of heat—enough to boil water continuously—thus violating the law of caloric conservation. Thompson suggested that the mechanical effect of friction was the true origin of heat. Sir Humphry Davy, the distinguished British chemist and popular science lecturer, helped Thompson to confirm the overthrow of caloric when he generated heat by rubbing two pieces of ice together on a cold London day. Heat, these researchers realized, is actually another form of energy—the ability to do work.

In spite of this simple and dramatic demonstration, the caloric theory persisted in pedagogic circles until the mid-nineteenth century. Scientists, however, had long since pushed ahead with the new interpretation. The English physicist James Prescott Joule, for instance, formalized Thompson's ideas by measuring precise transfers of mechanical energy to heat. Perhaps his best-known experiment involved measuring the change in temperature of water agitated by a paddle wheel. A sensitive thermometer registered a rise in water temperature that corresponded to the frictional energy expended by the turning wheel. Joule was thus able to establish a quantitative equivalence between an expenditure of mechanical energy and an increase in heat. In other words, Joule proved that heat is nothing more than a kind of energy.

THE LAWS OF THERMODYNAMICS

The recognition of heat as a form of energy was one of the great discoveries of science. Prior to this revelation, neither heat nor energy could be understood in any comprehensive way. By 1850, just a few years after Joule's experiments, the underlying principles of energy science—thermodynamics—had been set down by Rudolf Julius Clausius. Clausius presented two statements, the first and second laws of thermodynamics, which unify the behavior of energy as Newton had done for forces and motion.[4]

Following Thompson's studies, many scientists independently recognized that energy is conserved—the first law of thermodynamics.[5] Energy, the ability to do work, comes in many forms. Potential energy can be stored in a battery, a coiled spring, a match, or a boulder perched at the edge of a cliff. Kinetic energy exists in every moving object, from orbiting planets and comets to a pitcher's fastball. Heat is yet another form of energy. The first law of thermodynamics states that while energy can be changed from one of these forms to another, the total amount of energy never changes. The stored chemical potential energy of gasoline is converted to the explosive energy of expanding gas in a car's engine. That energy is transferred to the kinetic energy of the car when the driver accelerates, or it is dissipated as engine and brake heat when the driver engages the brake. Clock pendulums, bouncing balls, and life itself all depend on the constant transfer of energy from one form to another. But no matter how many times it is shifted, the energy is always there in one form or another.

The second law of thermodynamics places a severe restriction on the kinds of energy transfers that are possible. There are many equivalent statements of the second law, but perhaps the most cogent for a society dependent on fire was this: energy always converts from more concentrated

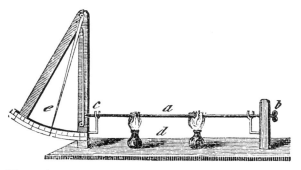

Thermal expansion of a metal rod could be measured in the laboratory with a pyrometer. Such measurements were critical in developing the science of thermodynamics. Illustration from John Lee Comstock, *Elements of Chemistry* (1853).

(that is, more useful) to less concentrated (less useful) forms. Concentrated fuel sources like coal, oil, and natural gas are valuable because they provide the best energy source and because they are limited resources. Once burned, the energy stored in coal or gas still exists, but in the much less useful form of dispersed heat. This hard fact of nature was all too obvious to Americans who had to rely on dwindling supplies of wood or unreliable winter supplies of coal. The scandalously high price of coal in New York City during the winter of 1831–1832 was a result, in essence, of the second law of thermodynamics.[6]

In its formal mathematical guise, the second law of thermodynamics leads to several important consequences. Some of these conclusions conform to everyday experience: for example, heat always flows from hot to cold. The second of law of thermodynamics does not preclude heating homes in winter or cooling them in summer. It merely says that if you want to shift heat "against the current" of hot to cold, then you must pay the price in energy.

The flow of heat from hot to cold provides a firm theoretical basis for the concept of temperature and the creation of temperature scales. Temperature and heat are different quantities. Two objects are at the same temperature if no heat flows spontaneously between them, no matter what their relative size. Heat, on the other hand, depends on the amount and type of material in question. Two gallons of water at 25 degrees centigrade contain twice the heat energy of one gallon at the same temperature. Furthermore, a gallon of water at room temperature contains several times the heat energy of a similar weight of metal, because the heat capacity (the heat required to raise a material's temperature a given amount) is much greater for water.

Temperature scales are defined arbitrarily by identifying two easily re-produced temperatures, such as the melting point of ice, the boiling point of water, or an exact volume of mercury in a narrow glass tube. Many different temperature scales were proposed in the eighteenth and nine-teenth centuries. The Dutch instrument maker Daniel Gabriel Fahren-heit, for instance, calibrated his mercury thermometers by defining ice water as 32 degrees and human body temperature as 96 degrees. Subse-quent refinements of the Fahrenheit scale used 32 degrees and 212 de-grees as the freezing and boiling points of water, respectively. In 1742 the Swedish astronomer Anders Celsius created the first scale based on one hundred divisions by defining ice water and boiling water as 0 degrees and 100 degrees, respectively. The Kelvin temperature scale has the same cen-tigrade divisions but defines 0 as "absolute zero," the complete absence of heat, which corresponds to about –273.15 degrees on the Celsius scale.[7]

The second law of thermodynamics has implications for the efficiency of engines, that is, devices that convert energy into work. According to the second law, it is impossible to build an engine whose only effect is to convert heat into an equivalent amount of work. In other words, there are no perfectly efficient machines because some heat energy is always lost to the environment. But the law in all its mathematical rigor does much more than simply repudiate the possiblity of perpetual motion machines. It can be used to calculate the optimal efficiency for any type of engine, whether powered by steam, gasoline, or electricity. In an industrial society where fuel savings meant greater profits there was a tremendous incentive to de-sign the most efficient engine, and the second law of thermodynamics pointed the way for many improvements.

Finally, the second law suggests a profound truth about order in the universe and the directionality of time. The shift of energy from more concentrated to less concentrated forms is a one-way street, leading any closed system to an increasingly disordered condition. An iron bar, cool at one end and red hot at the other, is in a relatively ordered condition, with thermal states well separated end to end. Over time, however, those or-dered states will become randomized over the bar as heat shifts from the hot end to the cool end. Nineteenth-century philosophers saw the iron bar as a microcosm of the universe, which will eventually cool to a uniform, lifeless state—the "heat death" of the universe. Time is an arrow pointing toward this inevitable end.[8]

THE KINETIC THEORY OF HEAT

From the recognition of heat as a form of energy, it was a relatively small step to deduce that heat energy is actually a manifestation of moving atoms—kinetic energy at the atomic scale. Materials at high temperature contain rapidly moving or vibrating atoms, whereas cooler substances have less atomic motion. This idea seems almost intuitive to us now, but it was a bold and abstract concept to nineteenth-century researchers, who had no direct evidence for the existence of atoms. The kinetic theory of matter was developed during the third quarter of the nineteenth century by a number of leading physicists, including Clausius, James Clerk Maxwell, Ludwig Eduard Boltzmann, and John Tyndall.[9]

The kinetic theory of heat provided a framework for research into other physical phenomena. Albert Einstein's first major theoretical work, published in 1905, used the kinetic theory to explain Brownian motion, the random jittering of small dust and pollen particles suspended in liquids.[10] Einstein developed a quantitative model for the magnitude of the phenomena, including its increase in activity with increasing temperature. The work provided a compelling confirmation of the kinetic theory, as well as the first widely accepted experimental evidence for the atomic theory of matter.

The Nature of Flame

Heat is inseparable from flame, but understanding flame requires much more than the laws of thermodynamics or the kinetic theory. Ancient theories of fire treated flame as a substance, one of the four classic elements. This concept was well entrenched by the late seventeenth century, when Georg Ernst Stahl presented his phlogiston theory, a somewhat more sophisticated version of the fire-as-substance hypothesis.[11]

THE PHLOGISTON THEORY

Stahl provided perhaps the first unified, systematic approach to chemical reactions, although he did so without benefit of an atomic theory. In keeping with the Aristotelian concept of flame, he defined phlogiston as the aspect of any combustible material that is lost during burning. In terms of this definition, phlogiston would be the principal component of good fuels. According to Stahl, such fuels as natural gas, anthracite coal, and whale oil (all of which burn virtually without residue) are essentially pure phlogiston, while wood, peat, and bituminous coal contain phlogiston plus ash.

Stahl used this model to explain rusting, respiration, and fermentation, as well as most types of combustion. As fires burn, phlogiston is released into the atmosphere. Flames are extinguished in a closed chamber when the air becomes saturated with phlogiston. The opposite effect occurs in the reduction of ores to metal, claimed Stahl. Metal contains more phlogiston than its ore, as expressed by the reaction:

$$\text{metal ore} + \text{phlogiston (i.e., charcoal)} \rightarrow \text{metal}.$$

By the same token, burning a material like phosphorus produces both a phosphorus ash and phlogiston:

$$phosphorus \rightarrow phosphorus\ ash + phlogiston.$$

Careful measurements of the products and reactants of these chemical reactions revealed a surprising and ultimately unacceptable property of phlogiston. Metal ore weighs *more* than metal plus phlogiston; dephlogisticated phosphorus ash weighs *more* than phosphorus. In other words, phlogiston seemed to have negative mass. The discovery of hydrogen, the "inflammable air," deflected some of the objections for a time because this lighter-than-air gas seemed to be a pure form of phlogiston. The reprieve was brief, however, and the demise of the phlogiston theory came at last with the discovery of oxygen.

THE DISCOVERY OF OXYGEN

The phlogiston theory was destroyed by one of its staunchest supporters, Joseph Priestley. Prior to Priestley's landmark studies of the 1760s and 1770s, only three different gases ("airs") were recognized by scientists: inflammable air (hydrogen), fixed air (carbon dioxide), and the earth's atmosphere. To this list Priestley added ten more, including his most famous discovery: the gas we know as oxygen.[12] Candles burned with such intensity in the presence of this gas that Priestley called it "dephlogisticated air"—air completely unsaturated with phlogiston and therefore capable of sustaining the most vigorous flame. He claimed that this air was effectively the chemical opposite of phlogiston.

Antoine-Laurent Lavoisier, the brilliant French chemist, seized on Priestley's experimental discoveries and quickly incorporated them into a new theoretical framework that facilitated a more quantitative approach to chemical reactions.[13] Lavoisier challenged the phlogiston theory by arguing that, when burned, sulfur and phosphorus increase in weight by the addition of air, while metal ores are converted to pure metals by the loss of air. When Lavoisier first proposed this idea in 1770, the detailed nature of the "air" in question was unknown, but Priestley's discovery of dephlogisticated air in 1774 provided a key piece of the puzzle.

Lavoisier's concept of chemical reactions helped him to recognize the singular role of this new air, which he called oxygen. In his view, phlogiston was a complete fiction. It was oxygen that combined with combustible materials like sulfur, charcoal, or phosphorus in the process of burning. Thus, the phosphorus reaction should be expressed as

$$phosphorus + oxygen \rightarrow phosphorus\ ash.$$

Lavoisier also demonstrated that burning coal produces another well-known gas, namely, "fixed air" or carbon dioxide. The reaction for the reduction of metal ore (including common ores of lead, tin, and iron) could therefore be expressed as

$$\text{metal ore} + \text{charcoal} \rightarrow \text{metal} + \text{fixed air.}$$

The power of Lavoisier's approach to chemical reactions was not merely its mode of expressing products and reactants. This new theory also had the enormous advantage of placing chemistry on a firm quantitative basis. The weight of products was found always to be equal to the weight of reactants: in chemical reactions mass is conserved, and there was no need to resort to negative weights. A given volume of oxygen, carefully prepared and measured, always reacts with a given amount of charcoal to produce a predictable volume of carbon dioxide. Similarly, that volume of oxygen reacts with phosphorus to produce a predictable quantity of phosphorus ash. Lavoisier's work led to the realization that the earth's atmosphere is a mixture of different gases. He demonstrated that water is produced by burning inflammable air (hydrogen), he proposed the existence of elements (chemicals that could not be separated into simpler substances), and he was able to deduce the compositions of many simple compounds.

Lavoisier's life was brought to a violent end in 1794 during the Reign of Terror. By then, however, he had already established a sound chemical basis for understanding flame. The next generation of chemists—notably John Dalton, who quantified certain aspects of the atomic theory, and Sir Humphry Davy, who discovered many new elements and compounds—applied Lavoisier's quantitative approach to various chemical reactions and further refined our understanding of the nature of combustion.[14]

Not surprisingly, this vast amount of research led to a fairly simple explanation of fire. By the 1820s combustion was finally understood by scientists to be a heat-initiated chemical reaction that results in the rapid combination of oxygen with other chemicals. This reaction releases energy in the form of more heat, along with light in most instances. The three essential ingredients of all fires are heat, oxygen, and fuel. Remove any one of these and the fire goes out.

Fire users would not have been particularly surprised at this latter statement. They had seen it happen all too often.

ANATOMY OF A CANDLE FLAME

For more than two centuries, researchers have used the flame of a candle as the focus for intellectual inquiry. As early as the 1730s, for example, Benjamin Franklin posed the question "Why does a candle flame burn upwards?" to members of the Junto or Leather Apron Club in Philadel-phia. Although Franklin could not have hoped for sophisticated answers, the query was an early contribution to a body of data that would increase dramatically over time. Indeed, scientists study candle flames today and continue to make startling discoveries.

By the second quarter of the nineteenth century, the basic chemistry of a burning candle had become clear.[15] The fuel of an unlit candle consists primarily of solid hydrocarbons, that is, compounds built principally from carbon and hydrogen atoms. These atoms are organized into strongly bonded groups called molecules, typically with about twenty carbons and twice as many hydrogens in the paraffin of the best candles. Chemical bonding between hydrocarbon molecules is weak, which accounts for the softness of candle wax. A slight increase in temperature to only about 50 degrees centigrade causes the paraffin to melt and form the familiar puddles and drips. Liquid wax also soaks into and saturates the wick. At higher temperatures, above 300 or 400 degrees centigrade, wax vaporizes into gaseous molecules of paraffin and smaller molecular fragments. This transformation of fuel into a combustible form, a process called pyrolysis, yields hot hydrocarbon vapor that readily reacts with oxygen molecules of air.

Close examination of a candle flame reveals a complex structure, more or less radially symmetrical around the wick. Much of the action takes place in the thin bluish wall, a boundary layer with temperatures close to 1400 degrees centigrade that defines the envelope of flame. This main reaction zone is where atmospheric oxygen, in the form of O_2 molecules, combines with the hydrocarbon gas to form carbon dioxide, water, and heat. Ideally, the reaction can proceed as follows:

$$3O_2 + 2CH_2 \rightarrow 2CO_2 + 2H_2O + \text{heat.}$$

In reality, though, many other reactions are also involved. Closer to the center of the flame, where temperatures are about 800 degrees centigrade, hydrocarbons can separate into carbon and hydrogen atoms. Hydrogen burns to water vapor, while the carbon atoms form into clumps of soot. Most of the bright yellow flame of the candle arises from an incandescent 1200-degree cone of this carbon soot. Those glowing carbon particles com-bine with oxygen near the tip of the flame to produce more carbon dioxide and heat.

Apart from heat, light is the most obvious attribute of the candle flame. The nature of light was not at all clear to early nineteenth-century observ-ers. Heat and light seemed to be different entities because a dying fire loses its glow before it loses its heat. Yet there is also an obvious relationship between the temperature of a fire and the brightness of its flame. It was not until James Clerk Maxwell's brilliant synthesis of electromagnetic phenomena in 1861 that the related nature of all kinds of radiation finally became clear.[16] Maxwell's electromagnetic equations led logically to the recognition of a wide range of wavelike phenomena, all of which have the characteristic speed of 186,000 miles per second. Visible light is but one small part of the great electromagnetic energy spectrum. Infrared radiation, or heat, was seen to be just like light, though slightly less energetic and thus produced at somewhat lower temperatures. Every candle, therefore, produces a spectrum of electromagnetic radiation. Some of it is visible, and much of it—at longer infrared wavelengths—is not.

The average candle user of the last century had scant interest in invisible phenomena. Of much more concern were the highly visible and problematic fires that cooked food, heated homes, and powered industry. Scientists had a lot to say about these fires, too, particularly in relation to the performance of various fuels. Although such investigations were often motivated by practical concerns, this in no way compromised the importance of the information derived. On the contrary. Science was valued by most educated Americans of the last century not only because it was a noble academic pursuit but also because it could be used to point the way toward new products and processes. Nowhere, perhaps, was this philosophy more fully realized than in the creation and development of new fuels.

THE SCIENCE OF FUELS

Oxygen and heat sustained America's fires, but the third requirement—fuel—provided an enormous amount of flexibility. Americans during the nineteenth century could choose from dozens of carbon-based fuels, each with distinctive properties, uses, and modes of pyrolysis. Fire research carried out prior to 1900 focused heavily on the question of fuels and, over time, served to characterize and promote a diverse array of solid, liquid, and gaseous materials suitable for combustion.[17]

Wood and Charcoal

The "science" of wood burning is as old as fire itself. Empirical knowledge of selecting, drying, and arranging wood for fires long preceded any attempts to quantify the phenomenon; in a sense, every fire user performed an experiment each time a fire was built and sustained. Given the insights of Lavoisier and his fellow chemists, however, it was easy to apply more rigorous methods to improving fuels.

The many varieties of dried wood fuels differ considerably in their residual water content, density, ash (noncombustible solid) content, bushels of charcoal per cord of wood, and heat produced per unit weight. Researchers analyzed thousands of trees and burned countless cords of wood in an effort to compare the many varieties available to Americans. Hickory and oak proved to be especially dense and efficient fuels, whereas maple, pine, and poplar were shown to be much less concentrated fuel sources.[18]

Wood passes through several stages of pyrolysis and combustion, and this fact led to a thriving charcoaling industry.[19] All wood contains water in the form of isolated H_2O molecules; as much as 50 percent by weight of freshly cut poplar is water, for example. When a log is placed on a fire, it begins a pre-ignition phase as these water molecules boil off. Dried wood is composed primarily of cellulose, long molecules containing chains of carbon and oxygen plus hydrogen. Cellulose pyrolysis involves the breakdown of these long chains into shorter carbon-rich segments that combine to form a carbon-rich solid (char) plus smoke containing volatile organic molecules such as turpene, tar, and resins. This everyday charring process occurs when a person burns a piece of toast, singes a shirt while ironing, or smokes a cigarette. As the wood ignites, char and tar smoke follow different physical and chemical pathways. Volatiles burst into the bright, dancing flame characteristic of wood fires. Solid carbon-rich char, on the other hand, produces the glowing combustion characteristic of hot embers or coal.

For many day-to-day applications Americans wanted high heat with a minimum of flame. Most natural fuels release volatile hydrocarbons and thus burn with a flame. The distinctive pyrolysis of wood, however, facilitates a gradual charring without burning if the wood is heated in an oxygen-poor environment. Volatiles are thus removed, leaving a carbon-rich fuel that we call charcoal.[20]

As mentioned in Chapter 3, the preparation of charcoal was a delicate task, requiring constant vigilance on the part of a skilled collier. Generally the process was accomplished in hemispherical mounds from ten to sixty feet in diameter. Carefully stacked lengths of wood were covered with tree limbs, leaves, and an insulating layer of dust. The first critical step, the "sweating" process, involved subjecting the wood to sufficient heat to dry, but not enough to cause an explosion from too rapid release of the yellowish-gray vapor. As the color of smoke shifted to light gray, the collier adjusted holes that provided just the right amount of ventilation to the mound. The time required to char a heap depended on the season, the weather, and the wetness of the wood. According to Frederick Overman, who described the process in 1850, it usually took three or four days to char a heap measuring thirty feet in diameter. Four to five days more were re-

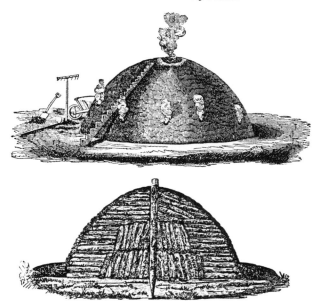

Wood can be converted to charcoal by gradual heating in the absence of oxygen. In 1850 most charcoaling was accomplished in mounds about thirty feet in diameter. From Frederick Overman, *The Manufacture of Iron* (1850).

quired to cool the mound, after which the charcoal could be raked and transferred to storage houses. Later in the century charcoal ovens, which were less subject to the vagaries of weather, replaced the simpler mounds.

Coal and Coke

It would be difficult to overestimate the impact of coal on nineteenth-century American society. Although the United States was slower to adopt coal than Great Britain, by 1850 coal had become an integral part of the lives of tens of thousands of Americans, particularly residents of the Northeast. At a time when wood was a diminishing resource, the mineral fuel was used to illuminate cities, heat homes, power locomotives, and refine ores. The adoption of coal was hardly an overnight phenomenon, however. Coal burning mandated entirely new technologies for effective use in both home and industry. Thirty years of experiment, invention, and publicity were needed to develop the requisite apparatus and to convince the American people that coal was a safe and reliable source of energy.

In the 1830s, about the same time that coal was being widely introduced in American homes, American scientists were beginning to recognize the origins of the nation's vast coal beds. Although a few observers suggested a purely mineral origin of coal, most geologists pointed to the abundant plant remains in coal formations as evidence for coal's "vegetable origin."[21] Coal deposits, they realized, are the metamorphosed remains of ancient plant debris, usually from bogs and swamps. During metamorphism, the organic deposit is gradually enriched in carbon as volatiles, including water, carbon dioxide, and hydrocarbons, are released. In a sense, therefore, coal formation is the natural analog to charcoaling. Considerable variation is seen in the composition and combustion properties of different coals, owing both to the nature of the original flora and to the stage of chemical alteration that has been attained.

By the 1840s few scientists debated the organic origins of coal, but the time frame for its formation was still a matter of contention. Advocates of a literal biblical interpretation saw coal as the clear consequence of a great recent flood, while many scientists viewed coal as the remains of ancient plant life. Aspects of this debate persist today in conflicts between biblical creationists and those who support the theory of gradual evolution.

In the nineteenth century, however, coal users, and the miners who supplied them, had much more pragmatic concerns. They wanted to know how good the coal was and how much it was worth. Miners evaluated the quality of a coal seam primarily on the basis of its "rank"—that is, the percentage of fixed carbon available for burning—and on the nature of its impurities, notably sulfur, which could contribute to a corrosive flame.

All gradations of fuel exist, from unaltered organic matter such as wood and leaves, through peat and lignite, to soft and hard coals. Nineteenth-century researchers quantified their chemical analyses of coal into three parts: carbon, volatiles, and ash. Formal classification of coal was based on the relative proportion of these components, as well as on the behavior of the coal under combustion. Deposits with less than about 50 percent carbon, including peat and lignite ("brown coal"), were not generally employed as fuel in America. The two principal commercial varieties of coal were bituminous, with 50 to 80 percent carbon, and anthracite, with more than 80 percent carbon.[22]

Bituminous coal, also known as "pit coal," "sea coal," or "soft coal," contains various impurities, which often lead to a sooty flame and significant quantities of ash after combustion.[23] The most common variety of bituminous coal is "fat coal," which is rich in resins and bitumen and often tends to swell and cake in a fire; particles of the coal may become cemented together into a hard mass under heat. Fat bituminous coal often required a hot-air blast in a confined chamber in order to burn efficiently; it was thus referred to as "close-burning coal." "Dry" and "very dry" bituminous coals, on the other hand, have fewer of these impurities and burn in an open fire. They were referred to as "open-burning."

Bituminous coal was the only mineral fuel available to most American homeowners prior to the 1830s. Soft coals were applied to domestic use as early as the mid-eighteenth century, especially in Virginia near the Richmond coalfield. Other soft coal localities, including western Pennsylvania and the Cumberland Valley, were also sites of early coal use. In addition, soft coal was imported in modest quantities from Britain, Nova Scotia, and Virginia by cities on the East Coast, and by 1800 it was a common, if not plentiful, commodity. Wood was generally favored for home use, however, because bituminous coal produced a great deal of sooty smoke, which was widely condemned by physicians and homemakers alike. At a time when most meals were prepared over an open fire, cooks avoided the possibly deleterious effects of coal-cooked food.

The presumed danger of stored coal was another factor in homeowners' reluctance to use fossil fuel. It was commonly believed that bituminous coal, if improperly ventilated, could undergo spontaneous combustion. The gradual seepage of gases from stored coal was also cited as a threat. Concerns about these possible health and safety hazards in the home were not fully resolved until the second half of the nineteenth century.

An important step in the widespread adoption of coal for industrial production of heat was the introduction of coking prior to combustion. Coking coal is exactly analogous to charcoaling wood, and the techniques are very similar. Coking could be accomplished in heaps or rows a few feet high or in coking ovens, which could handle several tons of coal at a time. Coke could be used to advantage at any facility that employed charcoal. Coke-fired furnaces were used for smelting ores, burning lime, boiling salt brines, manufacturing glass, and many other enterprises. Coke also proved excellent for steam generation at a time when steam engines were finding increased applications in American mines, mills, factories, and transportation devices.

The most highly metamorphosed coals are called anthracite, or "hard coal," and may reach 95 percent carbon content. Anthracites are hard, brittle, and burn with a smokeless flame—the ideal domestic fuel. It was clean to handle, clean burning, compact, and inexpensive relative to wood and softer coals. Nevertheless, anthracite was not immediately adopted, owing to confusion about its proper use. Most grates for wood and soft coal elevated the fuel and thus provided a substantial underdraft of cool air to maintain proper combustion. Anthracite, by contrast, required very high temperature to initiate combustion and so burned best with little or no

By the turn of the century Pennsylvania coke production was accomplished in rows of ovens near the bituminous coalfields. Stereograph by the Keystone View Company, ca. 1910.

draft on a low or sunken grate in which air could be heated. The introduction of Pennsylvania anthracite to eastern American markets in 1820 was followed over the next decades by dozens of patents for improvements in hard-coal stoves. Many of these ideas were based on scientific investigations described by such men as Professor Denison Olmsted of Yale University.[24]

By the mid-nineteenth century, anthracite was accepted as the most desirable domestic fuel. The new anthracite-burning stoves provided an improved source of heat and revolutionized America's culinary arts. Questions of health and safety were resolved as experience was gained and improved burning techniques were developed. An ever-growing, reliable supply of coal was ensured by an expanding network of canals and railroads. Thus, anthracite ushered in a new era of domestic comfort and contributed in no small way to the transportation improvements of the United States.

Oils, Fats, and Petroleum

Fuels for illumination must burn with a brilliant flame, an effect best achieved with volatile hydrocarbons. Among the dozens of fuels employed in nineteenth-century America for this purpose were vegetable oils (such as rape oil and olive oil), animal oils and fats (including tallow, stearine, and whale oil), and plant and bee waxes. Though quite different in source and consistency, all of these natural hydrocarbon-rich fuels are similar in chemical composition. Each releases gaseous molecules when heated and thus can be burned in a manner analogous to the candle to provide an ideal source of portable domestic illumination.[25]

The procedures for distilling flammable compounds from plant and animal tissues were often labor-intensive, and prior to 1800 domestic production satisfied most lighting fuel needs. Tallow candles, for example, were made from the solidified fat of sheep or cattle. Stringlike wicks could be coated with tallow by repeated dipping in the melted fat, or the melted fuel could be poured directly into candle molds. Improved candles composed of "spermaceti" (a whale-derived wax) replaced the cruder tallow-based forms in the 1830s, and those in turn gave way to cheaper stearine candles manufactured from a chemically purified animal fat or vegetable oil in the 1850s. The modern paraffin candle, a product of the petroleum industry, was developed about a decade later.

The solid fuels of candles were supplemented by a variety of liquid lamp fuels. Whale oil provided an ideal lamp fuel, though it was significantly more expensive than vegetable oils. As the price of whale products rose in the 1840s, much effort was expended on new designs for lamps that could effectively burn cheap vegetable oils, melted lard, or wood distillates such as turpentine or rosin oil.

The Canadian physician and geologist Abraham Gesner dramatically altered the course of lighting technology in 1846, when he discovered kerosene as a by-product of gas production from bituminous coal. This new lamp fuel was simple to produce and burned readily with a bright flame. Furthermore, Gesner realized, kerosene could be produced most easily from the abundant untapped supplies of petroleum that, until then, had been viewed as little more than curiosities. Petroleum refining began in earnest with the discovery of oil in western Pennsylvania in 1859. In little more than a decade, kerosene was the lamp fluid of choice in American homes.[26]

Petroleum is an extremely complex mixture of organic molecules—literally thousands of different molecules primarily composed of carbon and hydrogen—derived from prior life.[27] Every petroleum deposit contains a unique blend of these countless components. Some of petroleum's molecules are quite large, incorporating dozens of carbon atoms in complex rings and branching chains, whereas other molecules consist of just a few carbons in a simple row. The smallest of these molecules, with one to four carbons in a chain, vaporize quickly at room temperature. Propane (C_3H_8) and butane (C_4H_{10}), for example, are among the lightest molecules in petroleum. The remaining black liquid boils off in successive fractions as

By 1900 oil wells contributed to the growing economy of California. This oil field, located in Los Angeles, produced kerosene oil for local use and export to the Far East. Photograph published by the Philadelphia Museums, ca. 1900.

temperature is raised. Gasoline, for example, is the fraction of petroleum that vaporizes between 70 and 180 degrees centigrade and typically has five to eight carbons in its molecules. Owing to its high volatility, gasoline had little practical use before the widespread introduction of internal combustion engines early in the twentieth century. Kerosene represents a second stage of petroleum distillation, with a boiling range between about 180 and 270 degrees centigrade. Molecules with eleven to fifteen carbons, including both chain and ring forms, predominate in this fuel. At higher temperatures, distillates include gas oils (fifteen to twenty-five carbons), lubricating oils (twenty-six to fifty carbons), and a wide variety of solid residues, including asphalts, resins, paraffin waxes, and bitumens.

The nomenclature of nineteenth-century illuminating fluids is almost as complex as their chemistry. In addition to the vast number of specific molecular species, each with its own name, that contribute to such fuels, hundreds of varietal terms and brand names arose to describe the products of different companies. Furthermore, a given method of distillation, even if meticulously repeated from batch to batch, might produce widely varying results if applied to the petroleum from wells of different composition. The uniformity of modern petroleum-based products is achieved by synthesis; almost all of the gasoline we use, for example, is synthesized at refineries by "cracking" larger molecules into the desired fragments.

Coal Gas, Water Gas, and Natural Gas

During the eighteenth century, chemists in Britain and elsewhere discovered that volatile organic compounds, particularly methane (CH_4) and hydrogen (H_2) molecules, are released from coal upon heating. This phenomenon prompted widespread experimentation in laboratories across Europe, and in time "coal gas" found its way out of the laboratory and onto the streets—quite literally—where it became a favored fuel for urban illumination.

A number of British and European tinkerers experimented with gaslight, but credit usually goes to William Murdoch, who set up a small experimental plant in Soho in 1795. London approved the installation of some indoor and street lighting early in the nineteenth century, and by the 1830s more than half of the city was illuminated by coal gas.[28]

American enterprise was ready to follow the overseas leads. The first American demonstration of gas lighting occurred in Philadelphia in August 1796, just one year after Murdoch's pilot plant. Other exhibitions followed, although they were primarily small-scale curiosities. In 1812, for example, Rhode Island resident David Melville attracted considerable attention by lighting his Newport home, street, and a nearby factory in Pawtucket with gas.[29] The first citywide use of gaslight in the United States was authorized in Baltimore on June 17, 1816, shortly after the Peale family had demonstrated gas lighting in their public museum. Rembrandt Peale was given permission to manufacture gas, lay pipe, and contract with the city for street lighting. Within a few years many other cities—Philadelphia, Boston, and New York included—had followed suit.[30]

The principles of coal gas production are simple.[31] Prior to the Civil War, most gas was manufactured by placing bituminous coal—preferably a variety rich in volatiles—in a retort heated by a wood, coke, or coal fire. As volatiles were released, they passed through piping to a cold condenser, or trap, where organic compounds such as tar, naptha, and other residues could precipitate. The remaining gas was purified by bubbling it through a mixture of lime and water or by passing it over powdered lime. The purified coal gas was stored in a gasometer, which maintained a steady pressure in the network of gas lines to streets and buildings.

During the second half of the nineteenth century this simple process was gradually replaced by another, more efficient procedure: the water gas method. Researchers found that a flammable gas can be produced in abundance by passing superheated steam—700 degrees centigrade or hotter—over coal. In the absence of air, the reaction proceeds:

$$C + H_2O + heat \rightarrow CO + H_2.$$

Gas lighting revolutionized urban development and safety. The Herz Boulevard
Street Light, a portable model, was introduced about 1900.

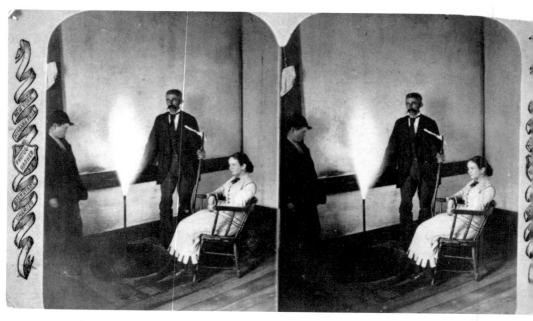

Natural gas was little more than a curiosity in 1875, about the time of this untitled stereograph
by Charles Bierstadt. The flaming pipe may be associated with the salt springs of the
Rochester, New York, area.

The resulting hydrogen-rich gas could be further "carburetted" (combined with carbon), to form a mixture of hydrogen and methane that burned with a bright, hot flame.[32]

The progress of the gaslight industry flickered in 1879, when Thomas Edison introduced his electric light bulb. The transformation to electric lighting was gradual, however, due in no small measure to the improved illumination provided by the Welsbach mantle. But by the end of the 1890s the age of electricity was overtaking the gaslight era. Rather than collapse in the face of such superior competition, however, the gas industry turned to heating technologies. As electricity inexorably transformed the way communities lit their homes and streets, gas heat claimed an ever-growing share of the huge home and industrial heating market.

Natural gas—composed primarily of methane (CH_4)—forms abundantly deep underground by the natural heating of coal within the earth. Gas wells were noted as American curiosities more than two centuries ago. George Washington designated a "burning spring" in Charleston, West Virginia, as part of a public park in 1775, and many sites in Pennsylvania, New York, Ohio, and other coal-rich areas boasted similar phenomena. A few small-scale local ventures sought to exploit this resource, but natural gas made virtually no impact on the coal gas industry until the 1920s, when improvements in drilling and long-distance piping transformed the process.

FIRE SCIENCE AND SOCIETY

To understand the process of combustion was to have the power to make better fuels and build better fires. Yet the great majority of Americans during the last century did not acquire—indeed, did not desire to acquire—such knowledge. Furthermore, many of those who did learn something about combustion focused only on such information as was needed for the operation of a particular piece of equipment. The anthracite stove provides a good example of this sort of selective learning. As mentioned earlier, Professor Denison Olmsted of Yale produced a pamphlet in 1837 with the intention of diffusing "among the more intelligent portions of our community, a knowledge of the *principles* on which the most successful management of coal fires depends." Olmsted expected his readers to relay this knowledge to others, even to "superintend, personally, the construction and regulation of their fires, until their domestics are furnished with the necessary practical skill."[33]

Just how many domestics were trained in this way is impossible to say. There were other ways of learning about fire science in the last century, however. Most of these forums were aimed at the educated and well-to-do, although some campaigns—the crusade for "scientific housekeeping," for instance—had wider exposure. Briefly, we will consider the formal and informal methods of educating the public about the fundamentals of fire science during the last century.

Formal Education

As soon as the Revolution ended, American intellectuals felt compelled to define the mission and content of a proper American education. Almost from the start, the sciences were accorded high esteem, not so much for their innate intellectual value as for their usefulness. Chemistry, botany, and geology were deemed especially important because mastery of these disciplines would help Americans unlock the riches of their vast continent and, in so doing, create a better world.[34]

The American expatriate Benjamin Thompson gave expression to this utilitarian philosophy of science in 1797 when he wrote to a colleague, "I can conceive of no delight like that of detecting and calling forth into action the hidden powers of nature! Of binding the Elements in chains, and delivering them over the willing slaves of Man!"[35] This was not idle talk. Thompson (Count Rumford) invented an enormous number of improved heating and cooking devices based on what he had learned from his scientific investigations of heat and the flame.

American schools and colleges gave expression to the philosophy as well. They incorporated chemistry, including the study of fire and heat, into their curricula with regularity and enthusiasm. Many of these institutions started their scholars young. As a survey of some of the more popular academic publications of the last century reveals, most juvenile textbooks treated oxidation, combustion, and the nature of candle flames in detail, and many included additional sections on the technology of steam engines and the applications of fire to domestic technology.

A good example of this approach is John Lee Comstock's popular *Elements of Chemistry*, which went through numerous editions during the 1830s and 1840s. The book commences with an extensive chapter on heat and caloric, including detailed information on temperature scales, thermometry, types of fuels, and the chemical reactions accompanying combustion.[36] Even more popular was Comstock's *System of Natural Philosophy*, first introduced in 1830 (the 1860 New York printing is designated "218th edition"). This textbook deals with such subjects as mechanics, hydraulics, and pneumatics, and in the process it describes a variety of engines, pumps, and other fire-related technologies.[37]

Also widely circulated, particularly in female academies, were the nu-

merous American editions of *Conversations on Chemistry*, by Jane Haldimand Marcet, an Englishwoman who had evidently attended the chemical
lectures of Sir Humphry Davy.[38] As stated in the original British edition of
1806, Marcet intended "to offer to the public, and particularly to the female sex, an Introduction to Chemistry." The prolific Comstock edited the
"tenth American from the eighth London edition, revised and corrected,"
in 1826, and many more editions appeared during the 1830s, 1840s, and
1850s. Chemical principles are presented in the classic catechism format.
Although these conversations may seem stilted, they clearly convey a vast
amount of information. A typical exchange early in the text relates to the
conduction of heat and features "Mrs. B." speaking with two children:

> *Mrs. B.* Free caloric always tends to diffuse itself equally; that is to say,
> when two bodies are of different temperatures, the warmer gradu-
> ally parts with its heat to the colder, till they are brought to the
> same temperature. . . .
>
> *Emily.* Cold, then, is nothing but a negative quality, simply implying
> the absence of heat.
>
> *Mrs. B.* Not the total absence, but a diminution of heat; for we know
> of no body in which some caloric may not be discovered.
>
> *Caroline.* But when I lay my hand on this marble table, I feel it *posi-
> tively* cold, and cannot conceive that there is any caloric in it.
>
> *Mrs. B.* The cold you experience consists in the loss of caloric that
> your hand sustains in an attempt to bring its temperature to an
> equilibrium with the marble. . . .[39]

Following a discussion of caloric, light, and the properties of several other
elements, Mrs. B. moves on to the chemistry of a burning candle:

> *Mrs. B.* In general, however, whenever you see flame, you may infer
> that it is owing to the formation and burning of hydrogen gas; for
> flame is the peculiar mode of burning hydrogen gas, which with
> only one or two apparent exceptions, does not belong to any
> other combustible.
>
> *Emily.* You astonish me! I understood flame was the caloric produced
> by the union of the two electricities, in all combustion whatever?
>
> *Mrs. B.* Your error proceeded from your vague and incorrect idea of
> flame; you have confounded it with light and caloric in gen-
> eral. . . .[40]

A footnote to this exchange modifies the original "hydrogen gas" to
"hydro-carbonated gas," in accordance with recent discoveries regarding
the composition of paraffin. Other corrections were added throughout the

century. It is debatable how much these conversations did to dispell the "vague and incorrect ideas" of many youthful chemists, but perhaps they stimulated a desire to learn more later in life.

If so, there were plenty of publications with which older students could pursue their studies. Chemistry textbooks by European and American experts appeared regularly. Friedrich Knapp's *Chemical Technology; or, Chemistry Applied to the Arts and to Manufactures* (1848) and Benjamin Silliman's *Elements of Chemistry* (1830) are good examples of such basic academic publications. The scientific principles contained in more challenging works such as James Clerk Maxwell's *Theory of Heat* (1870) and Ludwig Boltzmann's statistical mechanics were expressed in formidable mathematical terms, but these, too, found an appreciative readership among educated Americans, many of whom eagerly absorbed the new ideas about fire. Even though most of the important discoveries of the period had been made in Europe, the basic science was quickly translated for students on the other side of the Atlantic, and most college graduates would have had some exposure to the ideas current in their day.

What they chose to do with these ideas was another matter, of course. Most educators emphasized the utility of science. Walter R. Johnson, for instance, claimed in his preface to Knapp's *Chemical Technology* that chemistry had a direct bearing "on the welfare of every class, and every member of society."[41] It is likely, then, that some students went on to apply their book learning to real-life situations. For those attending school toward the end of the century, a number of educators made the connection for them. Most notable in this regard was the educational movement that sought to apply science to domestic life.

An early advocate of such an approach was Mrs. D. A. Lincoln, author of the *Boston School Kitchen Text-Book* (1883). She stated her philosophy at the outset: "It is as really a part of education to be able to blacken a stove . . . or to prepare a tempting meal of wholesome food, as it is to be able to solve a problem in geometry."[42] She then introduced readers to the science of cooking, describing the pertinent elements (oxygen, nitrogen, hydrogen, and carbon) and devoting an entire chapter to the management of fire.

Lincoln's primary concern was to apply the principles of science to cooking, but the notion clearly had merit for housekeeping as a whole. Indeed, by the turn of the century the movement for scientific housekeeping was well under way. Whereas only four American colleges had domestic science departments in 1890, that number had increased to twenty-one by 1899. At first, the instruction was directed primarily toward young working-class girls in order to prepare them to become better

···BOSTON···
SCHOOL KITCHEN
TEXT~BOOK

LESSONS·IN·COOKING

For the Use of Classes
in Public and
Industrial Schools.

❦ BY ❦

MRS·D·A·LINCOLN·

ROBERTS·BROTHERS· PVBLISHERS·
·· BOSTON ··

Mrs. D. A. Lincoln made sure that her students learned a little chemistry before they learned to cook.

household servants, but the emphasis gradually shifted to women of the middle class. By 1916, 195 institutions had almost 18,000 students studying home science and economics.[43] These programs were supplemented by other forums for instruction and self-improvement. Educator Ellen Swallow Richards formed the Boston Sanitary Science Club, the Boston School of Housekeeping, and the Women's Educational and Industrial Union—all for the purpose of teaching middle-class women how to use modern science and technology to improve the workings of the home. Many issues were addressed by these programs, but fire, because it figured so prominently in so many domestic operations, was a recurring theme.

Interestingly, the nation's firefighters, another group that could have benefited from fire science, did not pursue this option enthusiastically. As Rebecca Zurier has pointed out, firefighters were a notoriously conservative fraternity. Even though "fire colleges" (advanced training institutes for firemen) appeared with some frequency during the 1920s, it was not until after World War II that the power of science was applied vigorously to the problem of uncontrolled combustion.[44]

CHAPTER 6

Informal Education

For many nineteenth-century Americans, school was a luxury; for many more, course work beyond the rudiments was an impossibility. It is probable that fewer than half of all school-age white children attended public school in 1850. Although the statistics improved over the next fifty years, scientific literacy was hardly universal by the century's end.[45] Motivated adults could enhance their understanding of science in later life, however, either building on lessons learned in childhood or branching out into new territory.

Reading was a fundamental part of this process, and the textbooks mentioned above were often used by curious adults. Even more accessible were the general overviews of science that appeared regularly throughout the century. Insatiable American readers made best-sellers of works on natural philosophy and chemistry by the British chemist and popularizer Humphry Davy[46] and the American writer Samuel Griswold Goodrich (alias Peter Parley).[47] The public at large also bought more specialized accounts, such as Achille Cazin's *The Phenomena and Laws of Heat* (1869), a 265-page popularized version of John Tyndall's technical monograph *Heat Considered as a Mode of Motion* (1862).[48] Cazin describes the kinetic theory of heat on page 2, and the subsequent text draws heavily on both the science of heat and the everyday experiences thought to be shared by readers.

Encyclopedias and dictionaries of arts and manufactures were even more popular than these monographs. Often compressed in a single volume, they furnished short, easy-to-digest discussions on a vast array of topics. Multiple American editions of such works as Andrew Ure's *Dictionary of Chemistry* and *Dictionary of Arts, Manufactures, and Mines* and William Thomas Brande's *Dictionary of Science, Literature, and Art* provided the educated American with access to many details of fire science and technology.[49] These volumes had the added attraction of utility. All contained practical information on a variety of fire-related activities: instructions for setting up furnaces, forges, stoves, and lighting; recipes for making fireworks, colored glass, gunpowder, and steel; and step-by-step guidelines for bleaching, tanning, dying, and assaying gold.

Supplementing these comprehensive reference works was a steady stream of essays in such popular periodicals as *Scientific American*. At mid-century, every issue of this newspaper-format weekly contained features relating to fire science or technology, an emphasis that remained more or less constant to the end the century. Many of the articles were serious discussions of the latest discoveries in chemistry and physics, but these were often balanced by fascinating descriptions of fun-with-fire activites—how to produce a flame with ice, for instance, or how to make flames of different colors. A typical do-it-yourself fire project, "To Kindle a Fire Under Water," appeared in September 1845:

> Put into a deep wine-glass, that is small at the bottom, three or four bits of phosphorus, about the size of flax seeds, and two or three times the quantity of chlorate of potass, in grains or crystals, and fill the glass nearly full of water. Then place the end of a tobacco-pipe stem directly on, or over the chlorate and phosphorus, and pour nearly a tea-spoonful of sulphuric acid into the bowl of the pipe, that it may fall directly on the phosphorus; a violent action will ensue, and the phosphorus will burn vividly, with a very curious light under water.[50]

Scientific American also spiced up its more sober news of scientific discoveries and inventions with sensationalistic pieces on spontaneous combustion, explosions, arsonists, and destructive fires. Such an approach undoubtedly helped to sell copies, but these articles and their authors also heightened public awareness of the science of fire.

So did universities, lyceums, and museums, all of which sponsored educational programs for the general public throughout the nineteenth century. Among their diverse offerings, these institutions occasionally presented popular versions of the fundamentals of fire science and chemistry, often arranging for prominent scientists to speak and, if they were lucky, to perform intriguing experiments. Probably the most famous of these programs was the lecture series presented at the Royal Society in London. It was there that Sir Humphry Davy discoursed on heat and flame and performed such feats as melting ice by friction. Beginning in 1848, Davy's brilliant assistant and successor, Michael Faraday, continued the tradition and instituted an annual Christmastime lecture, "The Chemical History of a Candle." In Faraday's view, the candle was the perfect starting point for a scientific discourse. "There is no more open door by which you can enter into the study of natural philosophy than by considering the physical phenomena of a candle," he claimed. "There is not a law under which any part of this universe is governed which does not come into play, and is not touched upon, in this phenomenon."[51]

Yale chemistry professor Benjamin Silliman (1779–1864) apparently concurred with both Faraday's celebration of fire and his commitment to educating the public. Possibly inspired by Davy, whom he met in London in 1805, Silliman began his own series of public lectures on chemistry and geology in New Haven as early as 1808. From the 1830s through the 1850s Silliman traveled extensively, presenting both lectures and demonstrations to an enthusiastic public. Among his wide range of subjects, fire was a recurring theme, be it the fire of a meteor, the fire within the earth, or the role of heat in chemical reactions. At the age of seventy-eight, Silliman

LECTURE COURSE

OF

Prof. B. SILLIMAN, Sen.,

BEFORE THE

YOUNG MEN'S CHRISTIAN UNION,

OF BUFFALO,

AT

KREMLIN HALL

On the Evenings of Nov. 5th, 9th & 12th, 1857.

SUBJECTS—FIRE AND WATER.

THE LECTURES WILL BE ILLUSTRATED BY DRAWINGS.

Tickets for the Course, 50 Cents; Single Tickets, 25 Cts., to be had at the DOOR AND BOOK STORES.

BUFFALO:
C. B. YOUNG, PRINTER.
1857.

The renowned Yale professor Benjamin Silliman made a virtual career of presenting science to the public, and he often spoke on fire.

was still active. "Fire and Water" were his announced topics for a lecture course presented to the Young Men's Christian Union of Buffalo in November 1857.[52]

Silliman presented hundreds of popular lectures and in the process must have educated tens of thousands of people about fire. The number may seem relatively small, especially when compared to the cumulative subscription lists of popular journals like *Scientific American* and *Niles' Weekly Register*. Yet Silliman's efforts were duplicated time and again, in place after place, by a small army of science lecturers during the last century. Some of these speakers were local personages: Deacon Nehemiah Ball, for example, presented lectures on science to the inhabitants of Concord, Massachusetts, as part of the town's lyceum program. Some were college teachers who taught mini-courses in the natural sciences for various Chautauquas and other summer self-improvement programs. And some, like Charles Willson Peale and his sons, were flamboyant professionals who devised a never-ending stream of fire-based exhibitions for the museum-going public.

The impact of these educational programs was uneven and, on many occasions, probably negligible. Books and lectures enhanced the lives of many people but could not begin to provide the in-depth coverage of formal classwork. Nor could they be expected to reach more than a small percentage of the population. Still, it was a start.

SCIENCE AND THE FIRE-BASED SOCIETY

As we have seen, there were numerous ways in which the science of combustion affected American society as a whole during the nineteenth century. Conversely, society itself exercised influence on the workings of science, that is, on the very ideas of science. Nowhere, perhaps, is this phenomenon more obvious than in the debates of the last century about the age and composition of the earth. Mankind's reliance on fire as an energy source profoundly affected the way specialists thought about the earth's history.

Consider, for instance, the question of the earth's internal configuration. Intense fires deep underground were a recurring feature of earth models from ancient times through the turn of the twentieth century. According to these theories, the earth's interior was a searing region of perpetual combustion, primarily of coal and sulfur (brimstone). The eruption of volcanoes with their accompanying sulfurous stench, the increasing heat of mines with depth, and the occasional inextinguishable underground coal fires that burned for years provided compelling evidence for this model. Later theories, beginning with the *Principia philosophiae* (1644) of René Descartes, tended to downplay the role of internal flame but still incorporated more than enough heat to make sinners tremble.[53]

A comfortable alliance reigned between science and religion, inasmuch as both agreed that the earth's interior was a physical place of eternal fire. This conformity also made it easy for both camps to present their views to the general population. Anyone who used fire on a daily basis could understand the notion of unbearable heat. Whether one called the source of this heat "hell" or simply "the center of the earth" mattered little. In either case there was a satisfying symmetry: fires on the surface of the planet and fires deep underground.

By the mid-nineteenth century, fire in the earth was a factor in a new geological question, and once again science and religion had roles to play. This time, however, scientists and biblical scholars found themselves at odds, and the average person was left almost totally outside the debate. The problem was simple: to determine the age of the earth. The answer, though, was a long time coming.

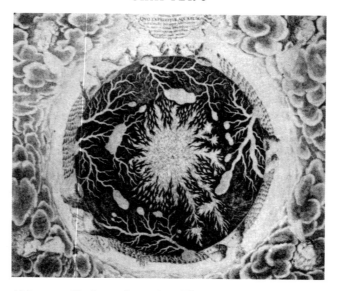

Althanasius Kircher, a Jesuit from Thüringin, proposed that the
earth's interior was composed of a network of subterranean "fire
chambers" interconnected by channels. His illustration and description, which appeared as part of *Mundus Subterraneus* (1678),
reflected the predominant view in nineteenth-century America.

Debate began in earnest when, around mid-century, the majority of
geologists adopted the uniformitarian view that geological processes occur
gradually and required immense spans of time to produce the features now
seen at the earth's surface. Estimates of an earth billions of years old were
perfectly consistent with the rock and fossil record, but this theory was
clearly at odds with a literal biblical interpretation that placed an upper
limit of perhaps ten thousand years on the planet's age. Even more disturbing was the implication that the earth had existed for hundreds of millions
of years before humans and that, therefore, the planet had not been created by God for the exclusive benefit of our species. Most geologists, however, persisted in their studies, analyzed their data, and ultimately rejected
the literal biblical interpretation in favor of the testimony of the rocks.

Then, unexpectedly, theoretical physicists rejected the testimony of the
geologists. British physicist William Thomson, later known as Lord Kelvin, tackled the question of the earth's age from a completely different
point of view. Thomson, a leader in the fledgling science of thermodynamics, realized that the earth and the sun are systems with limited energy
budgets. He posed questions about the total conceivable energy of the sun
and earth and how that energy must decrease with time as heat radiates

away from the cooling bodies. In a series of papers published over a thirty-year period, Thomson and his supporters argued that the earth and sun were losing heat at a rate much too fast to support a billion-year age estimate. His words were sobering:

> It is as certain that there is now less volcanic energy in the whole earth than there was a thousand years ago, as it is that there is less gunpowder in a "Monitor" after she has been seen to discharge shot and shell, whether at a nearly equable rate or not, for five hours without receiving fresh supplies, than there was at the beginning of the action. Yet this truth has been ignored or denied by many of the leading geologists of the present day, because they believe that the facts within their province do not demonstrate greater violence in ancient changes of the earth's surface, or do demonstrate a nearly equable action in all periods.[54]

Thomson and his colleagues based their analyses on a few logical assumptions. The earth's interior is hotter than the surface; therefore, energy is stored and generated from within. Heat constantly escapes by thermal conduction through rocks. The only significant sources of internal heat, apart from the residual heat of formation, are internal combustion and heat generated by the gradual contraction (and consequent conversion of gravitational potential energy) of a cooling globe. The sun, they assumed, is composed of an efficient combustible material like anthracite coal, and so it must be getting smaller as it burns. Thomson's published estimates of the time of crustal cooling (a prerequisite for life)—based on the best data available in 1862 on the earth's size, mass, and composition, the heat capacities and thermal conductivity of rocks, heats of combustion, and so on—was between twenty and four hundred million years.[55]

The upper estimate was acceptable to most geologists, although Thomson ruffled a few geological feathers by arguing strongly against uniformitarianism. The earth and sun, he asserted, must be cooling steadily, so past processes must have occurred under different conditions and at much higher rates. Supporters of Charles Darwin's theory of evolution were especially upset by Thomson's claims that as recently as a million years ago the sun was significantly hotter and larger and so produced much higher surface temperatures on earth. These conditions, Thomson asserted, must have produced violent storms, floods, and a very different pattern of vegetation.

As geologists scrambled to reconcile their field observations with Thomson's theoretical edicts, new data on rock and combustion properties allowed more refined calculations. The upper limit on the age of the earth's crust was reduced to one hundred million years, forcing earth scientists once again to squeeze the sweep of earth's history into an uncomfortably narrow box. In one sense, this result was viewed with excitement. Physicists could not support an age of more than one hundred million years; geologists could not accommodate a shorter span. The age of the earth seemed well defined.

This harmonious situation was disrupted by the American geologist Clarence King, a staunch advocate of Thomson's model. In 1893, while working for the United States Geological Survey, King undertook the most rigorous theoretical calculation yet of the earth's crustal age. According to King's seemingly unassailable calculations, the earth's crust, was a mere twenty-four million years old.[56] Thermodynamics and the science of combustion provided no leeway. Given King's data and assumptions, the oceans, sedimentary rocks, and life on earth could not have appeared until well after this date. How could this result be reconciled with the physical evidence of field geologists, many of whom would have preferred an age one hundred times greater?

Earth scientists had to wait until the first decade of the twentieth century, when physicist Ernest Rutherford recognized that the newly discovered phenomena of radioactivity was ubiquitous in rocks and thus provided the earth and the sun with another major energy source. Geologists and evolutionary biologists around the world must have breathed a collective sigh of relief at Rutherford's pronouncement: "The time during which the Earth has been at a temperature capable of supporting the presence of animal and vegetable life may be very much longer than the estimate made by Lord Kelvin from other data."[57]

With a sweep of his pen, Rutherford had dethroned combustion as the dominant universal energy source. Since before the dawn of recorded history, fire had symbolized nature's ultimate force for warmth and security, as well as for violent destruction and unbearable pain. Following the discoveries of Rutherford and his fellow physicists, scientists recognized another natural force—one vastly more powerful than the chemical reaction in its potential for good or evil. The nuclear age had begun.

7

Perpetuating the Flame

Scientific American covered the subject of fire from many angles in its issue for September 4, 1845. There were notices on fuel, patents for stoves, and a sobering account of the continent's fire losses that year. In addition, the editors included a rather lengthy sentimental poem, the evocative first stanza of which read:

> When burns the fireside brightest,
> Cheering the social breast?
> Where beats the fond heart lightest,
> Its humble hopes possessed?
> Where is the hour of sadness
> With meek-eyed patience borne?
> Worth more than those of gladness,
> Which mirth's gay cheeks adorn?
> Pleasure is marked with fleetness
> To those who ever roam,
> While grief itself has sweetness,
>   At home—sweet home![1]

Even in a scientific age, as this poem demonstrates, the old-fashioned open fire kept a powerful hold on the public consciousness. Despite the gradual technological transition away from the open flame during the course of the nineteenth century, potent forces worked against this process, virtually ensuring that open fires would be almost as familiar when the century ended as when it began. Generations of Americans kept fires in order to stay alive, but that is only part of the story. Some of them were also keepers of the flame in the sense of preserving traditional technologies amidst the swirl of progress.

COUNTERACTING THE TREND

The open flame persisted in part because of the extenuated character of the settlement process. During the long period of westward migration, rude camp fires—and the associated smoke- and ash-flavored food—were virtu-

This sentimental lithograph of a bachelor combines the hearth and an image of a loving family to suggest virtue. Peters Collection, National Museum of American History, Smithsonian Institution Photo 48672-C.

ally unavoidable. Thereafter, primitive fires continued to energize many aspects of domestic life. It was not unusual, for instance, for the construction of fireplaces and chimneys to take several years, during which time, as one midwesterner recalled, "the fire was made on the ground and the cooking done out of doors."[2] Moreover, while urbanites of the 1840s enjoyed the advantages of oil and gas lighting, migrants trekking west sometimes had to revert to the cruder techniques of earlier days. "Grease was put in an iron vessel in which was inserted a piece of wick or cotton, the outer end being lighted," was how one man remembered the illuminant of his pioneer childhood.[3] Although the mass distribution of stoves and kerosene lamps helped to reduce this recidivism somewhat, people who lived far from cities tended to live close to the open flame. Henry Adams's comparison between his family's up-to-date Boston house (which had a furnace and gaslight) and their outlying ancestral home at Quincy (which had an open fireplace, mantelpiece, and tinderbox) constitutes one man's small-scale experience of this large-scale historical pattern.[4]

Among certain occupations, too, reliance on the open flame was a fact of life during much of the century. Lumbermen, for instance, often worked in wilderness camps where open fires were commonplace. The stark description of the facilities at a New England lumber camp circa 1850 evokes the rudimentary conditions endured by American settlers of a much earlier period:

> They begin by clearing away a few of the surrounding trees, and building a log hut, which is roofed with bark, and provided with a cellar for lodging such of their goods as are liable to injury from the frost. The fire-place is in the middle of the hut, and the smoke goes out through a hole in the roof. Hay, straw, or branches of trees are spread on the ground, on which they all lie down at night to sleep, with their feet to the fire which is kept constantly alive. One person officiates as cook, whose duty it is to have breakfast ready before daylight, at which time all the party rise, and each take his "morning," or the indispensable dram of spirits before breakfast. . . . The whole winter is thus spent in unremitting labor.[5]

Cowboys, sheep ranchers, miners, field geologists, landscape painters, and soldiers are a few of the other occupational groups whose labors consistently took them to wilderness areas and who consequently had no choice but to use the open flame on a regular basis.

People engaged in more sedentary occupations also perpetuated outdated fire habits during the nineteenth century, although in these cases their behavior was voluntary and somewhat arbitrary. The emotional attachment to flickering flames was particularly strong among people of En-

Camping and campfires epitomized the nostalgia for the open flame in late nineteenth-century America. This photograph from Baraboo, Wisconsin, was taken in 1887.

glish descent during colonial days,[6] a fact that Benjamin Franklin took into consideration when he developed his stove with the open front. This penchant for fire watching survived the American Revolution with ease, and thereafter it flourished almost to the point of becoming a national obsession. Motivated variously by nostalgia for the good old days, by the belief that an open hearth represented family values and social stability, and by deep affection for the visual effect of fire, people of many backgrounds refused to allow open fires to disappear entirely from their lives.

It was easiest to achieve this goal on a part-time, recreational basis. Long before the advent of organized scouting and similar outdoor activities in the early 1900s, adults and children alike enjoyed informal forays into the countryside, where a day of adventure would inevitably be followed by an evening of relaxation at the campfire. Hunters presented one interpretation of this rustic life-style at their popular "Hunter's Camp" exhibit at the 1876 Centennial Exhibition. In subsequent decades, countless other groups devised their own versions. Bandsmen organized rural excursion parties, college students planned grand tours of the American West, fraternal associations went camping, and—much to the surprise of the English

visitor James Bryce—young men and women set off each autumn for overnight parties in the woods without benefit of chaperones.[7] Elinore Pruitt
Stewart's love of the mountains and wood fires was so great that she went
camping in the Wyoming wilderness with no one but her young daughter
for company. "Nothing could reach us on two sides," she wrote of one such
experience in September 1909. "In front two large trees had fallen so that
I could make a log heap which would give us warmth and make us safe."[8]

Stewart had no difficulty building a fire that night because she kept a
wood fire in a fireplace at home. In this she was unusual; but the true
measure of her originality was not her affection for the open hearth but
rather her dogged determination to maintain that tradition in the face of
technological change. Not that other homemakers had not tried. One
fairly effective technique was simply to resist adopting the newer fire-
hiding technologies when they appeared. Elizabeth Watters, for instance,
was ambivalent about her uncle's plans to install a Rumford oven in the
house. She wrote in 1803 that she was "pleased with my Uncles plan of
Rumfordising the mode of Cookery for the summer but agree with the
Cook that the sight of the Fire is far to be prefered in Winter."[9]

Close heating stoves met with similar objections when they were introduced, and for years it was more common to find such contrivances in
commercial establishments and large public areas than in the home. Not
surprisingly, this association delayed the introduction of stoves even
longer because they quickly acquired the reputation of being unsuitable
for domestic use. Even after the efficiency and suitability of stoves for individual households had been proven, however—and even after enterprising
manufacturers had thought to create parlor stoves with open grates or isin-
glass windows—advocates of the open flame continued their campaign,
sometimes quite colorfully. Albert Bolles reported a "feeling of unutterable
repugnance" at the sight of the new-fangled inventions and was convinced
that it was "perfectly impossible . . . to bring up a family around a stove."[10]
For him and his compatriots, the fireside remained a necessary feature of
home design.

Fortunately, plenty of architects and builders agreed with him. As
Lewis F. Allen explained in his *Rural Architecture* (1852), "A farm house
should never be built without an ample open fireplace in its kitchen and
other principally occupied rooms. . . . The great charm of the farmer's winter evening is the open fireside, with its cheerful blaze and glowing embers;
not wastefully expended but giving out that genial warmth and comfort
which, to those who are accustomed to its enjoyment, is a pleasure not
made up by any invention whatever."[11] Thirty years later J. Pickering
Putnam was still thinking along the same lines. He claimed categorically
that no "modern" house could be considered complete without a fireplace.

Even after technology had effectively removed
the need for the open flame in the home, the
tradition of the hearth remained strong. Flames
retain a central place in this fireplace design from
J. Pickering Putnam, *The Open Fire-Place
in All Ages* (1881).

Then, in an effort to update the concept, he published a variety of designs
for sophisticated fireplaces that were part of larger, housewide heating and
ventilation systems.[12]

Innovative as they were, Putnam's systems never really caught on. It
scarcely mattered, though. The mania for Colonial design, which had
seized the country at the time of the Centennial, persisted through the
1880s and 1890s and kept public affection for the fireplace strong. And
so it was to remain. The movement survived the cleanliness craze that
swept America between 1890 and 1900 in the wake of the frightening
revelations of bacteriological science. A decade later, it survived the minimalist period of architecture, which promoted a rational and restrained
approach to house design. It even survived World War I. As late as 1917,
Modern Priscilla magazine was still arguing in favor of fireplaces in the nurs-

An Absolutely Odorless "Home Fire"
The Latest Reznor Creation
Finished in Rich Japanese Bronze

No.	Height	Body Width	Extreme Width	Suitable for Opening	Crated Weight	Cu. Ft. per Hr. Consumption		List Price
						Natural	Artificial	
55	24″	18″	30″	20″ to 28″ wide	65 lbs.	2 to 20	4 to 30	$29.00
56	24″	22″	34″	26″ to 37″ wide	70 lbs.	4 to 30	6 to 40	30.00

This design (Adam Period) is one which will appeal to those who would replace the gas log with an efficient up-to-date appliance of the latest approved construction. Burns the Reznor yellow flame, the most cheerful of all gas fires, has all the advantages and none of the disadvantages of a blue flame. Once installed never needs repairs or adjustment, and always maintains a complete combustion of the gas. Burns with full efficiency under lower gas pressure than any other type of burner.

See pages 30 and 31 for detailed specifications.

For Natural or Manufactured Gas

The tips in burners of Reznor heaters cannot come out. (Patented feature)

The Reznor Manufacturing Company, Mercer, Pennsylvania, promoted gas-fueled inserts for the family fireplace. Their catalog, ca. 1910, illustrates a dozen styles of "home fires."

ery. "A fireplace is important for ventilation and for other reasons too," argued the author.

> We shall probably never cease to feel the deep influence which the hearth has had on mankind. Some one has said that you can paint a family group around a fireplace, but not around a steam radiator. The hearth is a natural gathering place where children watch the pictures in the leaping flame . . . [and where mother says] the quiet true things which are never forgotten.[13]

To many middle-class Americans it made no difference that these were emotional rather than scientific arguments—that even well-designed fireplaces were woefully inefficient as heat sources and that an open flame in the nursery was an invitation to disaster. Certainly the oft-repeated characterization of the family as an isolated unit was in direct conflict with economic reality. But, as its proponents would have readily admitted, the appeal of the hearth preservation movement was to the heart rather than the head; and the heartland of the country was evidently in tune with those who were intent upon retaining the fireside. There is no other way to explain the vast quantity of mother-at-the-hearth reminiscences published during the last quarter of the nineteenth century or to account for the postbellum mania for decorating—or, rather, enshrining—the fireplace with wooden shelving and other forms of ornamentation.

There certainly is no better way to justify the charmingly eccentric behavior of William Allen White's father, who, in an effort to recreate the rustic home of his youth, built a log cabin with a huge walk-in fireplace and then convinced his wife and child to forgo their up-to-date townhouse of the 1870s in order to live there. White recalled years later,

> My father's idea was to duplicate exactly the cabin in backwoods Ohio where he was born and grew up. He thought he could go back to the golden days of his youth, when Fear Perry used to carry a keg of cider into the house on her birthday, when John White wrestled with the Indians, when he and the boys—my father's brothers—cleared the land around Norwalk, in Huron County, Ohio. So there was a loft in our log cabin where the hired men slept. And hanging over the rafters, the first autumn, were strips of dried pumpkin, bunches of onions, tufts of sage, and Heaven knows what other nonsense groceries. I remember my father standing at a table, in a lean-to kitchen which we had built on the narrow end of the log cabin, cutting up beef to corn and to dry, and putting away pork to pickle. We had a smokehouse where we cured hams and bacon, and of course a pit where we buried cabbage, turnips, potatoes, carrots, beets, and apples.[14]

This romantic attempt to regain the past was sheer heaven to Willie, who slept in a trundle bed and kept his own pig. For his mother, however, it was an intolerable burden to have to cook on the open hearth and reproduce the exotic foodstuffs of an earlier era. Willie watched her reactions to the experiment with sympathetic interest. "Her indignation rose from mild, hooting, and fumbling attempts at sarcastic humor, where she was not highly competent, to the direct attack that always began somewhere in a high key with one phrase: 'Now I tell you, Doctor, this thing's just got to stop.'"[15]

The "thing" did stop for the Whites when, after about a year of country living, they moved back to town. For other segments of the population, however, it never even started. Lower-class families could afford neither fireplaces nor the luxury of exalting them. Radical feminists could afford them but wanted nothing to do with "bricks and mortar, dogs and drawing-room fenders" or anything else that represented the oppression of women.[16] Many practical-minded villagers across the country just could not understand the fireplace mystique. When writer Hamlin Garland wanted an old-fashioned fireplace in his house in West Salem, Wisconsin, during the 1890s, he had trouble finding a local builder who knew how to construct one.[17]

Even to some members of the middle class, the exaltation of the fireside began to seem stuffy and dangerously artificial after a while. These men and women would not have called home "the enemy of woman," as W. L George did in 1913, but they did seem intent on allowing a breath of fresh air to permeate the parlor. As Ann Douglas and others have pointed out, this reassessment of the home front was part of a larger trend toward the "re-masculinization" of American society in reaction to the overwhelming "feminization" that had occurred earlier in the century.[18] This subtle shift in values, which began in the 1880s and continued well into the next century, manifested itself in the increase in violent sports, in the reappearance of a forceful and manly clergy, and in the glorification of the outdoor life that was so dear to Theodore Roosevelt. There was also a phenomenal surge of interest in camping at this time, which, symbolically at least, pitted campfire against home fire in the minds of some. By this time, though, the hearth-and-home movement was being assaulted by an even more powerful adversary: electricity.

The adoption of electricity throughout America was uneven, partly because of technical problems and partly because of unfavorable attitudes among the public. To help overcome the latter, proponents of electrical power tried to ease the transition by maintaining the nomenclature of earlier technologies. Thus, Thomas Edison sought an electric light with sixteen "candlepower," the same as a gas jet. He also encouraged his electric company to call lights "burners" and to send bills for light hours rather than kilowatts.[19] Similarly, the Allegheny County Light Company promoted its electric signs with a progressive, yet comfortably familiar, message: "We light Luna Park. What would it be without the dazzling brilliancy? And every sign is an eye burner."[20]

Perhaps these details helped soften the transition to a new energy source, but surely the public was not misled for long. Although fireplaces would continue to be built, candles continue to be sold, and coal, gas, and even wood fires continue to heat homes and cook food, the advent of electricity was nothing short of a revolution. Several years elapsed between the appearance of the first successful incandescent light bulb in 1879 and the adaptation of this energy form to practical schemes of illumination, heating, and motion. Several decades elapsed before electricity reached rural areas, and several more passed before America's farms were electrified on a large scale.

Nevertheless, electricity was the energy source of the future. Fire lovers might have been reassured to know that combustion figured in its generation, but that could hardly have made up for the fact that electricity itself was invisible, intangible, and invariably produced outside the home.

For Henry Adams, the connection between the two forms of power was almost incomprehensible. Exploring an exhibition in Paris in 1900, he studied both steam engines and dynamos; but, try as he would, "no more relation could he discover between the steam and the electric current than between the Cross and the cathedral."[21]

As Adams put it, he had left the sensual world behind. If he accepted electricity at all, he had to accept it on faith.

FIRE AND THE ARTS

Americans perpetuated the open flame in yet another way during the nineteenth century: through their creative productions. Fire, both destructive and benign, was a popular theme for artists working in a variety of mediums. And while the masters of the "high" arts produced works of enduring value, it is in the realm of popular culture that we see most clearly the extraordinary hold that fire had on society at large. Whether working in literature, music, or the visual arts, American artisans labored long and hard to keep the flame alive.

The Visual Arts

For artists who wanted scenes that were overpowering, dramatic, and grand there could be no better subject matter than a raging conflagration. Amateur painters across the country attempted to capture local fires, and some achieved startlingly compelling images. In addition to these dabblers, professional artists tackled the subject from a variety of angles. Albert Bierstadt, for instance, painted an ominous scene of a burning whale ship just after the Civil War, and Hudson River painter David Johnson produced his eerie *Fire on Black Mountain* a few years after that. Prairie fires attracted the attention of a legion of painters, including William Ranney, Charles Wimar, Alfred Jacob Miller, and George Catlin. Charles Deas's last major painting was a classic, featuring heroically proportioned settlers running fearfully before the blaze.

Big fires demanded big treatments, and painters were prepared to deliver. Huge canvasses were not uncommon in Benjamin West's day, and by the 1840s such visionaries as Henry Banvard and Henry Lewin had produced views of the Mississippi River that sprawled across hundreds of yards of canvas and had to be rolled out on enormous wooden scrolls for viewing. Around the same time, scene painter John Rowson Smith produced a huge panorama of the burning of Moscow. When a group of Chicagoans decided to commemorate their city's great fire in paint, therefore, they had only to raise the money for materials and hire the artists to recreate the scene. Or, to be more accurate, the scenes. The final creation, encompassing a staggering twenty thousand square feet, depicted a swirling mass of burning buildings, smoldering ruins, and an endangered, but intact, cityscape. The most spectacular and thrilling section, according to one eyewitness, was in the northern division, where "the surging flames are sweeping through . . . with indescribable fury; the air is filled with flying brands, sparks and cinders; the streets are veritable whirlwinds of fire, falling walls and crumbling buildings on every hand." This connoisseur found the entire composition magnificent, "grand and awful in the extreme."[22]

Paintings were displayed to the public in galleries, trade expositions, and, in the case of the *Chicago Fire*, a specially designed building. Even so, their audience was necessarily limited. Woodcut prints and copper plate engravings, by contrast, had wide distribution even in the eighteenth century. With the introduction of the lithographic printing process, which achieved a firm commercial foothold in the United States by the late 1820s, the pictorial print was on its way to becoming an icon of American culture. Finely executed and yet affordable, these images were not only familiar to middle-class Americans but constituted the art work that decorated their homes. And of the wide variety of images offered to the American public, those with fire themes remained among the most popular throughout much of the nineteenth century.

Nathaniel Currier forged his career with fire. Two of his earliest prints, both issued in 1835, capitalized on the public's interest in fires and firefighters by depicting the aftermath of the Merchant's Exchange fire in New York and the Planters Hotel collapse in New Orleans.[23] With the tragic burning of the steamboat *Lexington* in 1840, Currier again transformed fire into art, this time so quickly (it took only three days for Currier and the New York *Sun* to have this spectacular print available on the streets) that the use of lithography to illustrate current events soon became standard practice.

THE GREAT CONEMAUGH VALLEY DISASTER. FLOOD & FIRE AT JOHNSTOWN, PA.

The Kurz and Allison lithograph of the 1889 Johnstown flood dramatically juxtaposes raging water and flame with desperate, trapped women and children. Peters Collection, National Museum of American History, Smithsonian Institution Photo 48569.

It was this capacity for timeliness, along with an obvious suitability for dramatic coloring, that made Currier's prints of blazing buildings so popular. By the time James Merritt Ives joined the firm in 1852, conflagrations were featured regularly. In its half century of existence, Currier and Ives issued now-famous views of the great fires in Boston, Chicago, New York, Pittsburgh, and St. Louis, as well as occasional scenes of burning vessels and prairie fires.[24] They competed with numerous other American lithographers, some of whom hand colored their prints in an assembly-line fashion as Currier and Ives did, some of whom used a highly effective color printing process called chromolithography. In either case, the scene was delineated on a stone by an artist or by a craftsman using an artist's design. The process allowed ample opportunity for artistic license, a situation exploited to the full by the Chicago firm of Kurz and Allison in their Johnstown Flood print of 1889. In this seething, dramatic representation, scores of classically proportioned victims struggle against stylized torrents of water that are punctuated at well-balanced intervals by starbursts of smoke and flame. The verbal description, "Hundreds Roasted Alive at the Rail-Road Bridge," underscores the visual intensity of the scene. Although the artists had clearly enhanced reality, most contemporaries would have appreciated the effort—if only because the print gave them a chance to see the disaster for themselves.

Photographers could offer the same kind of immediacy. For the portrayal of fires, two forms of photography—the stereograph and the picture post-card—were especially popular with a public that seemed almost insatiable in its appetite for scenes of destruction. Both were mass produced, and both were inexpensive enough to be easily affordable.

A stereograph is a pair of photographs that, when viewed through a special device called a stereoscope, appears to be three-dimensional. Intro-duced into America from Europe in 1851, stereoscopy gained momentum during the 1850s and 1860s. By the 1870s several American companies, including the Kilburn Brothers of Littleton, New Hampshire, and Charles Pollock of Boston, were producing images at the rate of several thousand cards per day.[25] Variety in selection was the aim of most large publishers, and, beginning with the citywide fire in Portland, Maine, in July 1866, firms began to offer fire devastation scenes along with their standard do-mestic cards and foreign views. Although there are no known stereos of the great Peshtigo, Wisconsin, fire of 1871, there are hundreds of images of the legendary fires in Chicago, San Francisco, and Baltimore. Sets of scenes were part of the standard marketing plans for many publishers, and it was not unusual for some of the larger companies to issue as many as seventy-five or a hundred separate views of a single fire. In addition, innu-merable lesser fires were chronicled by local photographers and offered for sale as stereographs. Many publishers included extensive descriptions of the fires on the cards with the photos, and some enterprising companies enticed buyers with clever before-and-after shots of buildings consumed by fire.

Stereographs remained popular until well into the twentieth century. By that time they faced competition not simply from other forms of enter-tainment but also from another form of photography: the picture postcard. In its heyday, from about 1898 to 1916, the picture postcard was valued both as a souvenir and as a means of communication. Like the stereograph, it skillfully recorded the devastating effects of urban fires; and, as with the stereograph, such views attracted eager consumers. In fact, amateur pho-tographers could produce their own postcards of local fires simply by re-questing that their druggist develop the negative onto photosensitive card-board stock that bore the inscription "Post Card." On the other hand, commercial publishing houses could document the great fires with varied

CHAPTER 7

VIEWS OF THE
BURNT DISTRICT, BOSTON.
PUBLISHED BY
H. G. SMITH, PHOTOGRAPHER, STUDIO BUILDING

1. Corner of High and Summer, looking east.
2. Church Green.
3. Church Green, with Co. D, First Regiment.
4. Hartford and Erie Railroad Depot.
5. Corner of Summer and South, looking South West.
6. Looking up Summer Street, showing ruins cor. Summer and High.
7. South West Corner of Summer and Chauncy Streets.
8. Trinity Church.
9. Summer Street, near Washington.
10. R. H. Stearns' Store, Summer Street.
11. North East Corner of Summer and Hawley Streets.
12. Thorndike Hall Building.
13. On Summer, from Church Green to Washington Street.
14. Corner of Kingston and Summer Streets (S. W.).
15. New Devonshire Street, from Summer.
16. Firemen Digging for Remains on Summer Street.
17. North East Corner of Summer and Devonshire Streets.
18. Co. D, First Regiment, M. V. M.
19. Hose Eight, and Tent.
20. Opening Vault of Bank of North America.
21. North of Franklin Street, near Hawley.
22. Franklin Street, toward Washington.
23. North Side of Franklin Street, near Washington.
24. Ruins of 85 and 87 Franklin Street.
25. Cathedral Building.
26. South East Corner of Devonshire and Milk Streets.
27. Milk Street, Looking toward Washington.
28. Pfaff's, on Milk Street.
29. Rear of Transcript and other buildings.
30. Pearl Street, from Milk.
31. North East Corner of Milk and Congress Streets.
32. North West Corner of Congress and Water Streets.
33. South East Corner of Congress and Water Streets.
34. South East Corner of Milk and Washington Streets.
35. On Milk opposite New Post Office.
36. Pearl Street, from Milk.
37. North East Corner Congress and Water Streets.
38. Water, from above Congress.
39. Rear Entrance of Post Office.
40. Looking North on Washington from Summer.
41. Looking South on Washington from Franklin.
42. Corner of Franklin and Washington, North East.
43. North East Corner Water Street and Liberty Square.
44. North West Corner Water and Kilby Streets.
45. South West Corner Water and Kilby Streets.
46. North West Corner Lindall and Kilby Streets.
47. Lindall Street, from Kilby Street.
48. Liberty Square, Kilby Street.
49. Water Street, from Liberty Square.
50. Kilby Street, towards State Street, from Liberty Square.
51. East side of Kilby, near Liberty Square.
52. South West Corner Milk and Oliver Streets.
53. Oliver Street, from Milk Street.
54. North West Corner Oliver and Sturgis, Pearl Street in the distance.
55. From Oliver, Corner Sturgis Street, looking South West.
56. Wendell, from Oliver, looking West.
57. Oliver and Wendell, looking South West.
58. Pearl and Sturgis, looking South West.
59. Sturgis and Oliver Streets, from Pearl Street.
60. West of Pearl Street, near Sturgis Street.
61. North of High Street, between Pearl and Oliver Streets.
62. Between Pearl and Oliver Streets.
63. East side Pearl Street near Wendell Street.
64. Wendell Street, from Pearl, with Hose 4.
65. Corner Pearl and Wendell Streets.
66. East side Pearl Street, near High Street.
67. West side Pearl Street, from Sturgis, toward Milk Street.
68. Pearl Street, from Sturgis Street.
69. 43 and 45 Federal Street.
70. Shoe and Leather Association Vaults on High Street, near Pearl Street.
71. Federal Street, from High Street, looking North.
72. North East Corner High and Federal Streets.
73. South East Corner High and Federal Streets.
74. North, from 221 Federal Street.
75. South West, from 221 Federal Street.
76. North East, from 221 Federal Street.
77. H. & E. R. R. Depot, and Burning Coal Wharf.
78. Near View of Burning Coal.
79. General View from 221 Federal Street.
80. Group on Corner High and Federal Streets.

Boston photographer H. G. Smith offered a set of
eighty different stereographs following the
1872 fire.

views and, because they used a photolithographic process, in enormous
batches. Many firms, such as Rieder, Cardinell and Company of Los Angeles and Oakland, produced numbered series of particular fires. If printed
in color, as was Britton and Rey's spectacular view of the "Complete Destruction of Cliff House By Fire," the sense of disaster could be heightened.

228

"A Fire Department in Action," a stereocard issued by the Keystone View Company, ca. 1900, shows a small-town company at work. This is the card that provided the brief history of fire mentioned at the beginning of Chapter 2.

The crowd tracks the progress of the flames at Cliff House in San Francisco, 1907.

Disaster did not appeal to everyone, of course. For consumers with a penchant for the gentler side of fire, artists created a wealth of appealing decorative items that romanticized fire as the symbol of domestic stability and security. Even in the most modern turn-of-the-century homes, where the open flame was safely insulated from the user, these paintings, lithographs, and other images allowed Americans to keep the flame alive. Among the most evocative of these works were the paintings and prints by genre artists; Enoch Wood Perry, Lilly Martin Spencer, and others portrayed fireside gatherings with happy families bathed in the hearth's ruddy glow. Eastman Johnson's immensely popular *The Boyhood of Lincoln*, an 1868 portrait of a youthful Abraham Lincoln reading by hearth light, captured the same mood in a historical context. Similar acclaim greeted the exterior domestic scenes painted by George Henry Durrie, who produced numerous images for reproduction by Currier and Ives. His snowy winter scenes were usually embellished with a smoking chimney and ample woodpile, symbolic of the warmth and security of home. Candles, both lit and extinguished, were incorporated as symbolic elements in still-life paintings by John Frederick Peto, William M. Harnett, John Haberle, and other artists of the late nineteenth century.

Almost any kind of fire fueled the imagination of artists and thus could grace the walls of homeowners. Winslow Homer depicted open-air recreational fires, James McNeill Whistler celebrated the colorful cascade of a fireworks display, and John F. Weir glorified the flaming forge in his epic industrial portrayal, *Forging the Shaft* (1877). From the blazing hunters' campfires of Arthur Fitzwilliam Tait's chromolithographs to the smoky industrial fires of Aaron Henry Gorson's impressionistic canvases, flame added drama to art.

Music

Although music is less easily identified with a particular subject than other art forms, it is not impossible to make a fire connection. Programmatic pieces were enormously popular toward the end of the nineteenth century, and, as mentioned earlier, they often glorified the exploits of firemen. They could also be composed to explore other fire themes. One such piece, a favorite of Pennsylvania's Allentown Band, depicted Napoleon's invasion of Russia in 1812—complete with the burning of Moscow. Less comprehensive but enormously popular with bands of the antebellum era was "The Wood-Up Quick Step," a sprightly tune that evoked the loading of fuel onto a steamboat. Other musical compositions with fire-based titles include "Wild Fire Quickstep" and "Fire Spirits." Although the title does not mention the hearth specifically, the connection is implied in "Home

Sweet Home," a staple of band programming during the nineteenth century. Printed programs sometimes referred to the piece as "our National fireside hymn."[26]

It was much easier to convey an idea through music if words accompanied the score, and, not surprisingly, plenty of songs of the period relate to fire. "The Two Orphans," by P. J. Downey, chronicled the tragic effects of the Brooklyn Theatre fire. In a similar vein "The Iroquois on Fire," by Zella Evans, served as a tribute to the 602 people who died in the worst theater fire in American history. Other songs with fire references were less lugubrious and alluded to such themes as love, the home, and soldiers in camp.

The Written Word

Almost any sort of literature could be enhanced by the use of fire. In the hands of the best writers, the device was both dramatic and symbolic. James Fenimore Coooper used raging fires in both *The Pioneers* and *Home as Found* as opportunities for character analysis as well as for riveting storytelling. In *The House of the Seven Gables* Nathaniel Hawthorne gives the enormous central chimney all the significance of a character. Edith Wharton also employed fire and its symbolism to advantage, particularly in *Ethan Frome*.

It is in the lighter fiction of the day, however, that the period's sentimental attachment to fire is most obvious. Poetry offered an especially effective medium, as the following anonymous verses exalting coal demonstrate.

THE COAL AND THE DIAMOND

A coal was hid beneath the grate,
    ('Tis often modest merit's fate;)
    'Twas small, and so perhaps forgotten;
Whilst in the room and near of size,
    In a fine basket lined with cotton,
In pomp and state a diamond lies.
    "So, little gentleman in black,"
The brilliant spark in anger cried,
    "I hear, in philosophic clack,
Our families are close allied:
    But know the splendor of my hue,
Excelled by nothing in existence,
    Should teach such little folks as you
To keep a more respectful distance."

At these reflections on his name,
The coal soon reddened to a flame:
Of his own real use aware,
He only answered with a sneer;
I scorn your taunts, good Bishop Blaze,
    And envy not your charms divine;
For know I boast a double praise,
As I can *warm* as well as shine.[27]

The following four stanzas are part of a longer poem by Jesse E. Dow that glorifies fire in the context of iron manufacture, which in the late eighteenth and early nineteenth centuries was often accomplished in rural furnaces.

THE IRON MASTER

I delve in the mountain's dark recess,
And build my fires in the wilderness;
The red rock crumbles beneath my blast,
While the tall trees tremble and stand aghast;
At the midnight hour my furnace glows,
And the liquid ore in a red stream flows
Till the mountain's heart is melted down,
And seared by fire is its sylvan crown.

Old Cyclops worked in his cavern dire,
To [illeg.] the arrows of Jove with fire;
I in my mountainside crevice toil,
Amd make the rocks in my cauldron boil,
That man may hurl on his fiercest foes,
The iron rain and the sabre blows;
And send on the long and quivering wire
The silent thought, with a wing of fire.

I burn the woods, and I melt the hills,
While the liquid ore from the earth distills,
That over the railroad track may run,
The iron horse to outstrip the Sun;
That ponderous wheels may dash the brine,
And play with monsters of the Line;
While isles of coral seem to be,
But mile-stones placed in the deep blue sea.

When night comes on and the storm is out,
And the rain falls merrily about,

My mountain fires with ruddier glow,
Are seen to burn by the drones below;
And as my merry men pass around,
Their shadows seem on the bright back-ground,
Each like a Vulcan huge and dire,
Forging a thunderbolt of fire.[28]

Fire also figured in narratives about railroads, miners, and sailors at sea, but nowhere was it so lovingly described as in domestic tales. There are endless examples of this genre, although few surpass Harriet Beecher Stowe's evocation of the family fireside in *Oldtown Folks* (1869). Her description focuses on a Sunday evening gathering in her grandmother's kitchen. The enticing qualities of the flame-warmed room stand in stark contrast to the chilly outdoors and the cold formality of the parlor.[29]

> We had come home from our second Sunday service. Our evening meal of smoking brown bread and baked beans had been discussed, and the supper-things washed and put out of sight. There was an uneasy, chill moaning and groaning out of doors, showing the coming up of an autumn storm, —just enough chill and wind to make the brightness of a social hearth desirable, —and my grandfather had built one of his most methodical and splendid fires.
> The wide, ample depth of the chimney was aglow in all its cavernous length with the warm leaping light that burst out in lively jets and spirits from every rift and chasm. The great black crane that swung over it, with its multiplicity of pot-hooks and trammels, seemed to

This early American woodcut depicts reading near an open fireplace.
From John G. Rogers, *Specimen of Printing Types from the Boston Stereotype Foundry* (Boston: S. M. Dickinson, 1832).

have a sort of dusky illumination, like that of old Caesar's black, shining face, as he sat on his block of wood in the deep recess of the farther corner, with his hands on the knees of his Sunday pantaloons, gazing lovingly into the blaze with all the devotion of a fire-worshipper. . . .

On the part of Aunt Lois, however, there began to be manifested unequivocal symptoms that it was her will and pleasure to have us all leave our warm fireside and establish ourselves in the best room, —for we had a best room, else wherefore were we on tea-drinking terms with the high aristocracy of Oldtown? We had our best room, and kept it as cold, as uninviting and stately, as devoid of human light or warmth, as the most fashionable shut-up parlor of modern days. It had the tallest and brightest pair of brass andirons conceivable, and a shovel and tongs to match, that were so heavy that the mere lifting them was work enough, without doing anything with them. It had also a bright-varnished mahogany tea-table, over which was a looking-glass in a gilt frame, with a row of little architectural balls on it; which looking-glass was always kept shrouded in white muslin at all seasons of the year on account of a tradition that flies might be expected to attack it for one or two weeks in summer. But truth compels me to state, that I never saw or heard of a fly whose heart could endure Aunt Lois's parlor. It was so dark, so cold, so still, that all that frisky, buzzing race, who delight in air and sunshine, universally deserted and seceded from it; yet the looking-glass, and occasionally the fire-irons, were rigorously shrouded, as if desperate attacks might any moment be expected.

Now the kitchen was my grandmother's own room. In one corner of it stood a round table with her favorite books, her great workbasket, and by it a rickety rocking-chair, the bottom of which was of ingenious domestic manufacture, being in fact made by interwoven strips of former coats and pantaloons of the home circle; but a most comfortable and easy seat it made. My grandfather had also a large splint-bottomed arm-chair, with rockers to it, in which he swung luxuriously in the corner of the great fireplace. By the side of its ample blaze we sat down to our family meals, and afterwards, while grandmother and Aunt Lois washed up the tea-things, we all sat and chatted by the firelight. Now it was a fact that nobody liked to sit in the best room. In the kitchen each member of the family had established unto him or her self some little pet private snuggery, some chair or stool, some individual nook, —forbidden to gentility, but dear to the ungenteel natural heart, —that we looked back to regretfully when we were banished to the colder regions of the best room. . . .

Things being thus, when my Uncle Bill saw Aunt Lois take up some coals on a shovel, and look towards the best-room door, he came and laid his hand on hers directly, with, "Now, Lois, what are you going to do?"

"Going to make up a fire in the best room."

"Now, Lois, I protest. You're not going to do any such thing. Hang grandeur and all that.

> 'Mid pleasures and palaces though we may roam,
> Be it ever so humble, there's no place like home.'

you know; and home means right here by mother's kitchen-fire, where she and father sit, and want to sit. . . ."

Ghost stories and love stories, daring escapades and moral tales—all used fire for subject matter, for atmosphere, and for romance. The creators of these tales, like the citizens around them, were unable to perpetuate fire's traditional roles forever. Through their writings, however, they preserved some sense of what those traditional roles meant to fire users of the last century.

Epilogue

Standing on the brink of the twentieth century, Henry Adams thought he detected the emergence of a new breed of American. He was, in Adams's words, "the child of steam and the brother of the dynamo." He and his fellow Americans had already achieved such wealth and visibility by the time of the St. Louis Exposition of 1904 that Adams dared to hope that they might, over the next fifty years, actually discover what it was they wanted. "Possibly," Adams mused, "they might even have learned how to reach it."[1]

What these "servants of the powerhouse" learned or did not learn on their journey through the new century is a matter of debate, but it is undeniable that their heritage—including their potent and problematical fire heritage—continued to shape their destiny. When a turn-of-the-century stereographer declared that "we think we could not live without matches and fires," he was absolutely correct.[2] Indeed, fire remained a reliable energy source throughout the 1900s and retains much of its utility and appeal today.

Combustion still constitutes the basic energy source for millions of American households. Although gas and oil are now the most common domestic fuels, the traditional wood fire has made a limited comeback, particularly among householders concerned about renewable resources. Electricity has supplanted the open flame for most of the nation's lighting needs and for countless other tasks as well, but it is not a fire-free form of energy. Most electricity is generated by plants that are fueled by the burning of coal and is thus, indirectly at least, a product of combustion.

Many modern industrial processes continue to rely on fire, as does one of the basic symbols of modern life: the automobile. Although in this case the flame is hidden from the user, the process is comfortably familiar: fuel (gasoline) is ignited by spark plugs, and hot expanding gases—primarily carbon dioxide and water—drive the pistons of the car. Indeed, combustion drives all manner of modern transport, from diesel engines to jet planes. The launching of a rocket is almost by definition a public celebration of the power of the flame.

Gouts of flame add much to the excitement of a space shuttle launch. Even in the age of electricity and nuclear power, fire holds a fascination. Courtesy of NASA.

No cord. No flame. But still our wonderful candle lamp provides that soft romantic light you cherish so on tabletop, mantel and nightstand. Batteries hidden inside the realistic taper power a flame-shaped clear glass bulb to create a warm, safe glow wherever and whenever one is needed. A Colonial-style candlestick of hand-cast, hand-polished solid brass completes the illusion. Boxed; batteries not included. The **taller lamp** (9" high overall), #1-16878-0. 12.00. The **shorter**, #1-16879-8. 10.00.

Signatures
19465 Brennan Avenue
Perris, CA 92379

S-12619R-1

Flickering electric candles, with their flame-shaped
bulbs and brass bases, attempt to capture the
traditional elegance of the open flame.

Amid these profoundly utilitarian views of fire, contemporary Americans sustain a sentimental attachment to the flame that their nineteenth-century forebears would have found admirable. Modern consumers buy electric lights in the form of candles, and they furnish their Colonial-style houses with reproductions of whale oil lights, Argand student lamps, and tin lanterns. "No cord, No flame. But still our wonderful candle lamp provides that soft romantic light you cherish," claims one of the myriad companies that manufacture and market such sentimental artifacts today.

Despite their proven inefficiency as heating devices, fireplaces have also remained popular throughout the twentieth century. Bungalows, which were introduced just after the turn of the century as efficient and economical alternatives to traditional housing, almost always included fireplaces to entice buyers. Likewise, William Levitt's low-cost housing developments of the post–World War II era guaranteed that the eight hundred square feet of living space allotted to each unit included room for a hearth. As we enter the last decade of the twentieth century, most builders of middle-class homes perpetuate this tradition proudly; and it is not uncommon for the more expensive models to include several fireplaces, along with up-to-date facilities for open-flame cooking in the kitchen.

There is something elemental about fire, and modern Americans are no less receptive to its allure than were their ancestors. Fire metaphors abound in popular music, and any number of entertainments (rock concerts and circuses come immediately to mind) capitalize on the dramatic impact of real flames. Campfires have managed to retain their mystique, and romantic dinners virtually require the aura of candlelight. Sophistication in advertising is surely one of the most cherished aims of modern corporations; yet products as diverse as cars, beer, and blue jeans are routinely touted with the glowing imagery of an old-fashioned flame.

In extreme cases, contemporary fires can even recapture their primordial identity as the life-sustaining force. The adventure of five Costa Rican fishermen demonstrates this point especially well. Lost and powerless in the wake of a storm, these men sailed aimlessly around the Pacific Ocean for five months. They could catch enough food to eat, but it was the presence of a little Bic lighter on board that really saved them. Not only did the lighter allow them to cook their fish, but, far more important, it gave them hope. The lowly lighter quickly became their most prized possession. They cleaned it lovingly and observed the periodic lighting of the flame almost as a ceremonial occasion. "During the trip that lighter seemed like life itself for us," recalled one of the men after their rescue. "The moment when [we] flicked the lighter into flame became an important, hope-affirming ritual."[3]

This overview of modern fire practices would seem to imply that Americans have managed to harness the power of the flame without subjecting themselves to its evil side effects. Nothing could be farther from the truth. Fire remains a bad master, as any daily newspaper will attest. Americans still suffer staggering property losses from the ravages of the flame—according to one estimate, at twice the rate of all other industrialized nations.[4] Although improved construction techniques have eliminated the citywide conflagrations that were typical of the past, about five million separate household fires break out across the country every year and an estimated five thousand Americans die in the aftermath.[5]

Not surprisingly, there are some thirty-five thousand fire departments and as many as three million rescue workers occupied nationwide to save property and lives from incineration. Although the principal causes of fires have changed somewhat over time—most domestic fires today are caused by cigarettes and electrical malfunctions—human nature has changed very little. "People are irresponsible and careless about fire," claims a social worker who attempts to help children who set fires.[6] Crime statistics suggest that the problem runs far deeper than mere negligence, however. Arson, which increased at an alarming rate during the 1970s and 1980s, is the suspected cause of fully half of all urban fire losses. The perpetrators are as difficult to catch as ever, but the number of arrests—almost twenty thousand in 1985 alone—is a sobering reminder of America's bleak fire history.[7]

To make matters worse, modern Americans suffer not only from their own careless fire habits but also from the pernicious side effects of the fire practices of their ancestors. Decades of combustion have begun to cause problems that we are only beginning to comprehend. For at least a century, the very composition of the earth's atmosphere has been changing at a rate sufficient to alarm even conservative observers of the global environment.

The most obvious problem concerns the amount of carbon dioxide in the air. Carbon dioxide is an inevitable by-product of the combustion of all carbon-based fuels: coal, oil, natural gas, wood, and the myriad petroleum-based energy sources. Forest fires, animal respiration, and decay have always contributed to the natural abundance of atmospheric carbon dioxide, but prior to the Industrial Revolution these sources were more or less balanced by plant photosynthesis and other natural processes that convert carbon dioxide gas to other forms of carbon in the oceans and soil. Over the past century, however, the concentration of atmospheric carbon dioxide has increased dramatically, perhaps as much as 25 percent.

Carbon dioxide is not dangerous per se, for it is neither toxic nor polluting. Rather, carbon dioxide is one of the so-called greenhouse gases: it is transparent to visible light and so lets in the sun's energy to the earth's surface, which then warms up. But these gases are opaque to heat radiation, which is thus trapped near the earth's surface. As the concentration of carbon dioxide and other greenhouse gases increases, global temperatures may start to rise. Even an increase of just a few degrees in average global temperatures could have widespread, devastating, and largely unpredictable effects. Melting ice caps could raise ocean levels and inundate many coastal areas. Changing weather patterns could turn fertile farmland, such as the American Midwest, into desert.

Other consequences of the burning of fossil fuels are less cataclysmic but nevertheless have become cause for concern. The burning of coal pumps sulfur and nitrogen into the air—elements that react to form sulfuric and nitric acid in rain. Acid rain has caused widespread destruction of forest and lake ecosystems, where life is sensitive to the increased acidity of soil and water. Acid rain also causes the gradual erosion of building stones and has led to the permanent loss of many architectural treasures. Even more fundamental are the concerns of many modern chemists who rely on carbon-based fuels as the raw ingredients for an ever-expanding variety of new materials. Scientists craft plastics, adhesives, synthetic fibers, lubricants, and countless other products from chains and rings of carbon atoms. These atomic engineers believe that in the near future—perhaps by the year 2000—carbon-carbon bonds in coal, oil, and wood will become too valuable to burn up. They would argue that we have already consumed far too much of this valuable resource.

And yet, although the good servant/bad master opposition still applies to fire in modern life, much has changed about the way we handle and view fire. To some degree, the change is related to technological innovation. Firefighters now map forest fires with computers and employ a variety of new materials to extinguish flames as well as to protect themselves. On the home front, consumers can buy manufactured "logs" that ignite at the touch of a match and flammable charcoal briquettes that burst into flame without benefit of lighter fluid. Perpetuating the never-ending quest for novelty and nostalgia, the National Association of Home Builders offers a thoroughly updated version of the traditional family hearth: a gas fireplace that is open to view on three sides and can be "turned on" by remote control.

More significantly, however, Americans think about fire differently in the 1990s. It is impossible to pinpoint the year or even the decade in which this change occurred, since Americans eliminated the open flame from their daily lives at different times, depending upon where they lived and in what economic sector they operated. As late as 1932, for instance, almost one-third of all American homes still used candles and oil lamps for illumination, even though electricity had been introduced into urban areas decades earlier.[8] Nevertheless, fire-safety experts agree that, in general, the past two generations of Americans "relate to fire differently than earlier generations." This is hardly surprising. As a rule, we no longer have to manipulate fire on a daily basis and thus cannot possibly perceive the flame as the ever-present and highly dangerous ally that our ancestors knew it to be. One expert calls contemporary fire usage "indirect" and "almost inadvertent."[9] Indeed, for many people fire connotes nothing more than the occasional lighted candle on a birthday cake. The phenomenon that once regulated the very rhythm of daily life has become almost irrelevant.

For confirmation of this fact, one has only to consider a novel fire-safety program launched in Maryland. Hoping to teach children how to survive household fires, officials have devised a portable fire chamber to simulate

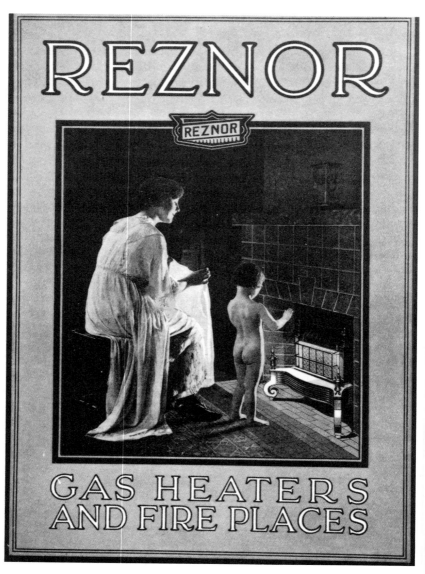

Gas heaters such as this one may have been put in the fireplace like a stack of wood, but they were in fact very different. The insulation of the user from the flame was taken one step further with the introduction of the "electric fire."

the real thing, smoke and all. It is a modern solution to an age-old problem, but its very existence underscores how much has changed. It is not simply that men and women of the nineteenth century would have been unable to control a fire in such a sophisticated way; they never needed to do so. Their children would have had ample opportunity to learn about the dangers of fire firsthand—from seeing fires in the home, from hearing about fires from relatives, and from following firemen as they went about their business.

In many ways, of course, our lives have been improved. We spend far less time on energy production and enjoy a greater degree of physical comfort. Sprinkler systems, fire alarms, and fire-retardant fabrics save property and lives from damage or destruction by the flame. Some observers have even suggested that the television has become our modern hearth and that this ubiquitous entertainment device draws people together in much the same way that fireplaces did in earlier times.[10]

These extraordinary developments would have startled such nineteenth-century fire lovers as the Beecher sisters, and the disappearance of the domestic hearth would surely have disturbed them. For them, as for so many of their contemporaries, fire was a "natural" energy source, as wholesome and as necessary as sunshine itself. Indeed, the comparison of hearth fires to the sun was often made outright. J. Pickering Putnam argued as late as the 1880s that people should heat their homes with open fireplaces because that mode of warming most closely approximated radiation from the sun. The Beecher sisters were more emotional in their analysis: "Warming by an open fire is nearest to the natural mode of the Creator, who heats the earth and its furniture by the great central fire of heaven, and sends cool breezes to our lungs."[11]

Here, in spite of themselves, these tradition-bound observers have hit upon yet another crucial difference between their world and ours in terms of energy production and use. Although the open fires of the nineteenth century looked and felt like the sun, they were, in fact, worlds apart. Whether fueled by wood, coal, gas, or oil, combustion is a simple chemical reaction. Solar energy, on the other hand, is generated by thermonuclear fusion reactions—the direct conversion of mass to energy, as immortalized in Einstein's most famous equation: $E = mc^2$. And, whether we like it or not, nuclear energy—with its promise of unparalleled benefits and its threat of unimaginable destruction—is an inescapable feature of our modern world.

Notes

1. Quoted in William Cronon, *Changes in the Land: Indians, Colonists, and the Ecology of New England* (New York: Hill and Wang, 1983), p. 25.

2. Quoted in James A. Ward, *Railroads and the Character of America, 1820–1887* (Knoxville: University of Tennessee Press, 1986), p. 30.

3. Philip Vickers Fithian, *Journal & Letters of Philip Vickers Fithian, 1773–1774: A Plantation Tutor of the Old Dominion*, ed. Hunter Dickinson Farish (Williamsburg, Va: Colonial Williamsburg, 1945), p. 255.

4. Fithian, *Journal*, p. 253; Arthur D. Pierce, *Iron in the Pines: The Story of New Jersey's Ghost Towns and Bog Iron* (New Brunswick, N.J.: Rutgers University Press, 1957), p. 103; Rachel Haskell, "A Literate Woman in the Mines: The Diary of Rachel Haskell," in Christiane Fischer, ed., *Let Them Speak for Themselves: Women in the American West, 1849–1900* (New York: Archon Books, 1977), pp. 58–72 (see p. 65); James Walter Goldthwait, *Diary of Western Trip, 1902* (Burnsville, N.C.: Ad Lib Press, 1982), p. 88.

5. Mrs. Orsemus Boyd, *Cavalry Life in Tent and Field* (New York: J. Selwin Tait and Sons, 1894), excerpted in Fischer, ed., *Let Them Speak for Themselves*, pp. 111–23 (see pp. 113–14).

6. *Phillips' United States Patent Fire Annihilator* (New York: Office of the Company, 1851), p. 3.

7. Thomas Bulfinch, *Bulfinch's Mythology* (New York: Avenel Books, 1978), p. 939.

8. Mark Twain, *Roughing It* (New York: New American Library, 1962), pp. 178–80, 139–40.

9. "Burning Yet," *Scientific American* 1, no. 14 (December 18, 1845).

10. Joseph Husband, *A Year in a Coal-Mine* (Boston and New York: Houghton Mifflin Company, 1911), pp. 72–171.

11. Mary Chesnut, *Mary Chesnut's Civil War*, ed. C. Vann Woodward (New Haven: Yale University Press, 1981), p. 800.

12. Benjamin Franklin, *An Account of the New Invented Pennsylvanian Fire-Places*, reprinted in Leonard W. Labaree et al., eds., *The Papers of Benjamin Franklin*, vol. 2 (New Haven: Yale University Press, 1959), p. 426.

13. Lucy Larcom, *A New England Girlhood, Outlined from Memory* (1889; rpt. Williamstown, Mass.: Corner House Publishers, 1977), pp. 21–22, 168, 23, 28.

14. Lincoln Steffens, *The Autobiography of Lincoln Steffens* (New York: Harcourt, Brace and Company, 1931), p. 53.

15. Elinore Pruitt Stewart, *Letters of a Woman Homesteader* (1914; rpt. Lincoln: University of Nebraska Press, 1961), pp. 216–17.

16. Walter Hough, *Fire as an Agent in Human Culture* (Washington, D.C.: Government Printing Office, 1926), p. 166.

17. Henry Adams. *The Education of Henry Adams: An Autobiography* (Boston and New York: Houghton Mifflin Company, 1918), p. 467.

18. Ibid., pp. 10, 466.

CHAPTER 2
GOOD SERVANT

1. "A Fire Department in Action," Keystone Stereograph P152 (18207) (Meadville, Pa: Keystone View Co., ca. 1900).

2. Archeologists in South Africa have uncovered fossilized animal bones thought to have been charred on hearth stones built anywhere from 1 million to 1.5 million years ago. C. K. Brain and A. Sillen, "Evidence from the Swartkrans Cave for the Earliest Use of Fire," *Nature* 336 (December 1, 1988): 464–66.

3. The studies referred to are Sir James George Frazer's classic investigation of magic and religion, *The Golden Bough* (1890–1915), and Walter Hough's monograph, *Fire as an Agent in Human Culture* (1926). Stephen J. Pyne's *Fire in America: A Cultural History of Wildland and Rural Fire* (Princeton: Princeton University Press, 1982) explores revealing regional differences in wilderness fire habits.

4. Abraham Lincoln's first inaugural address contains a passage that draws eloquently on this association: "The mystic chords of memory, stretching from every battlefield and patriot grave, to every living heart and hearthstone, all over this broad land, will yet swell the chorus of the Union."

5. These hints come from the following sources: white spots, Alexander V. Hamilton, *The Household Cyclopaedia of Practical Receipts and Daily Wants* (Springfield: W. J. Holland, 1875), p. 21; glass vessels and ants, Lydia Maria Child, *The American Frugal Housewife, Dedicated to Those Who Are Not Ashamed of Economy*, 12th ed. (1833; rpt. Cambridge, Mass.: Applewood Books, n.d.), pp. 20, 21.

6. Child, *American Frugal Housewife*, pp. 41–42.

7. Faye E. Dudden, *Serving Women: Household Service in Nineteenth-Century America* (Middletown, Conn.: Wesleyan University Press, 1983), p. 280.

8. John E. Brackett, "Training To Be Soldiers," in Marguerite Miller, ed., *Home Folks*, 2 vols. in 1 (Marceline, Mo.: Walsworth, n.d.), 2: 88–105.

9. Reference to this device, along with an amusing depiction, appears in James Burke, *The Day the Universe Changed* (Boston: Little, Brown and Company, 1985), p. 203.

10. Roger L. Welsch, comp., *A Treasury of Nebraska Pioneer Folklore* (Lincoln: University of Nebraska Press, 1966), pp. 370, 331, 345, 330, 356.

11. Walter Hough, *Fire as an Agent in Human Culture* (Washington, D.C.: Government Printing Office, 1926), p. 63.

12. Elizabeth F. Ellet, *The Pioneer Women of the West* (1852; rpt. Freeport, N.Y.: Books for Libraries Press, 1973), p. 43.

13. Richard Henry Dana Jr., *Two Years Before the Mast and Twenty-Four Years After* (New York: P. F. Collier and Son, 1969), p. 163.

14. An excellent analysis of American burning habits appears in Stephen J. Pyne, *Fire in America*. For a discussion of practices in colonial New England, see William Cronon, *Changes in the Land: Indians, Colonists, and the Ecology of New England* (New York: Hill and Wang, 1983).

15. John R. Stallard, "Soldiering Down South," in Miller, ed., *Home Folks*, 1: 134–46.

16. John A. Hickman, "Echo from the Pacific," in Miller, ed., *Home Folks*, 2: 8–15.

17. Quoted in Cronon, *Changes in the Land*, p. 48.

18. See Pyne, *Fire in America*, for an extensive discussion of this controversial question. The debate about the appropriateness of fire in wilderness areas raged for decades and, indeed, continues today, as the 1989 Yellowstone Park conflagration demonstrates.

19. Otis E. Young Jr., *Western Mining: An Informal Account of Precious-Metals Prospecting, Placering, Lode Mining, and Milling on the American Frontier from Spanish Times to 1893* (Norman: University of Oklahoma Press, 1970), p. 181.

20. This mode of ventilation was employed during the construction of the famous Petersburg Mine in 1864. See Jon D. Inners, "Colonel Henry Pleasants and the Military Geology of the Petersburg Mine, June-July, 1864," *Pennsylvania Geology* 20, no. 5 (1989): 3–10.

21. The first manned flight, undertaken by the Montgolfier brothers in 1783, featured a hot-air balloon heated by an on-board furnace that burned chopped straw. Because balloons of this design tended to deteriorate from the heat—and even burn—they were soon supplanted by reusable hydrogen balloons. During the 1950s, however, as a result of research by the Department of the Navy, reusable hot-air balloons were perfected and are again in operation.

22. Quoted in Jacob Bronowski, *The Ascent of Man* (Boston: Little, Brown and Company, 1973), p. 123.

23. James T. Hodge, quoted in Margaret Hindle Hazen and Robert M. Hazen, *Wealth Inexhaustible: A History of America's Mineral Industries to 1850* (New York: Van Nostrand Reinhold, 1985), p. 182.

24. Friedrich Knapp, *Chemical Technology; or, Chemistry Applied to the Arts and to Manufactures* (Philadelphia: Lea and Blanchard, 1848–1849), 1: 19.

25. Jesse E. Dow, "The Iron Master," *Scientific American* 1, no. 6 (October 2, 1845); Henry Wadsworth Longfellow, "The Village Blacksmith," in *The Poems of Henry Wadsworth Longfellow* (New York: Modern Library, n.d.), pp. 667–68.

26. Mary C. Henderson, *Theater in America: 200 Years of Plays, Players, and Productions* (New York: Harry N. Abrams, 1986), pp. 229–30.

27. Charles Coleman Sellers, *Mr. Peale's Museum: Charles Willson Peale and the First Popular Museum of Natural Science and Art* (New York: W. W. Norton and Company, 1980), pp. 228–31.

28. "Advance of Railway Science," *Scientific American* 1, no. 6 (October 2, 1845).

29. This description appears in Daniel J. Boorstin, *The Americans: The National Experience* (New York: Random House, 1965), p. 16. To cut slabs off a rock wall, quarrymen often built fires in front of the rock face and hoped that cleavage would result from the heat.

30. Lewis Binford has described these levels in some detail. See James Deetz, *In Small Things Forgotten: The Archaeology of Early American Life* (New York: Doubleday, 1977), pp. 50–51.

31. Mark Twain, *Roughing It* (New York: New American Library, 1962), pp. 139–40.

32. *History of Elkhart County, Indiana* . . . (Chicago: Chas. C. Chapman and Company, 1881), p. 155.

33. Quoted from an unidentified 1849 source in ibid., pp. 155–56.

34. Quoted in Ernest Earnest, *The Volunteer Fire Company, Past and Present* (New York: Stein and Day, 1979), p. 6.

35. Strong, quoted in Rebecca Zurier, *The American Firehouse: An Architectural and Social History* (New York: Abbeville Press, 1982), pp. 50–51 ("loaferage") and 51 ("hero of the piece").

36. Quoted in David D. Dana, *The Fireman: The Fire Departments of the United States, With a Full Account of All Large Fires, Statistics of Losses and Expenses, Theatres Destroyed by Fire, and Accidents, Anecdotes, and Incidents* (Boston: James French and Company, 1858), pp. 85, 65.

37. Basil Hall, *Travels in North America* (1829; rpt. Graz: Akademische Druck- und Verlagsanstalt, 1965), 1: 22; Zurier, *American Firehouse*, pp. 50–54.

38. Edward Everett Hale, *A New England Boyhood* (1893; rpt. Upper Saddle River, N.J.: Literature House, 1970), p. 169.

39. Phineas T. Barnum, *Struggles and Triumphs: or, Forty Years' Recollections of P. T. Barnum* (Hartford: J. B. Burr and Company, 1869), p. 702.

40. Harold Hancock, *The History of Nineteenth Century Laurel (Del.)* (Laurel: Laurel Historical Society, 1983), p. 337.

41. Hale, *New England Boyhood*, p. 166.

42. Quoted in Richard C. Wade, *The Urban Frontier: Pioneer Life in Early Pittsburgh, Cincinnati, Lexington, Louisville, and St. Louis* (Chicago: University of Chicago Press, 1959), p. 92.

43. H. W. Schwartz, *Bands of America* (Garden City, N.Y.: Doubleday, 1957), p. 27.

44. Ricky Jay, *Learned Pigs & Fireproof Women* (New York: Villard Books, 1987), p. 269.

45. Ibid., p. 242.

46. Ibid., p. 257.

47. Edward A. Barnwell, *The Red Demons; or Mysteries of Fire* (Chicago[?], 189[?]), 16 p. Author is listed as E. Barnello.

48. G.H.W. Bates & Company, *32d Annual Catalogue* (Boston: Bates & Company, ca. 1895), unpaged; see announcement for "The Fire Eater."

49. In authors' collection.

50. Or one-woman shows. Senora Josephine Girardelli is probably the best-known female perpetrator of fire tricks. An Italian, she arrived in England in 1814 and performed many times in London. According to Jay (*Learned Pigs*, p. 260), an American

"fire-queen" by the name of Miss Rogers also performed during the early nineteenth century.

51. Quoted in Irving Howe and Kenneth Libo, eds., *How We Lived: A Documentary History of Immigrant Jews in America, 1880–1930* (New York: Richard Marek Publishers, 1979), p. 243.

52. That staged fire was dangerous is undeniable. The most famous theater fire in history—the burning of Shakespeare's original Globe in 1613—resulted from the misfiring of a prop cannon.

53. Douglas Gilbert, *American Vaudeville: Its Life and Times* (1940; rpt. New York: Dover Publications, 1968), p. 30.

54. Frances Trollope, *Domestic Manners of the Americans*, ed. Donald Smalley (New York: Vintage Books, 1949), pp. 168, 172.

55. G. F. Patton, *A Practical Guide to the Arrangement of Band Music* . . . (Leipzig and New York: John F. Stratton and Company, 1875), p. 191.

56. George Thornton Edwards, *Music and Musicians of Maine* (New York: AMS Press, 1970), p. 333.

57. "Hope Valley, The Twin Villages' Jubilation," 1888, newspaper clipping in historical files of the Langworthy Public Library, Hope Valley, R.I.

58. For an excellent discussion of traditional European bonfires, see James George Frazer, *The Golden Bough: A Study in Magic and Religion* (New York: Macmillan, 1960).

59. For a sampling of such toys, see Bernard Barenholtz and Inez McClintock, *American Antique Toys: 1830–1900* (New York: Harry N. Abrams, 1980).

60. Flora Gill Jacobs, *Dolls' Houses in America: Historic Preservation in Miniature* (New York: Charles Scribner's Sons, 1974), pp. 62–65. A fire mark was a sign displayed on the exterior of a building to identify the insurance company that protected the property.

61. Sears, Roebuck and Company, *The 1902 Edition of the Sears Roebuck Catalogue* (New York: Crown Publishers, 1969), p. 912.

62. Hale, *New England Boyhood*, p. 118.

63. Sarah Bixby-Smith, *Adobe Days* (Cedar Rapids, Iowa: Torch Press, 1925), excerpted in Christiane Fischer, ed., *Let Them Speak for Themselves: Women in the American West, 1849–1900* (New York: Archon Books, 1977), pp. 247–59.

64. The symbolism of the birthday candle dates back to the Greeks and Romans, who used lighted tapers to communicate their desires to the gods. The use of lighted candles on a birthday cake is a German custom that perpetuates the belief that good things can come from invisible spirits through the conduit of a flame.

65. D. C. Beard, *The American Boys Handy Book: What To Do and How To Do It* (1890; rpt. Boston: David R. Godine, 1983), p. 373.

66. Lina and Adelia B. Beard, *The American Girls Handy Book: How To Amuse Yourself and Others* (1887; rpt. Rutland, Vt.: Charles E. Tuttle Company, 1969), pp. 196–98.

67. D. Beard, *American Boys Handy Book*, pp. 144–46.

68. Frank Dillon, "First Love Explained," in Miller, ed., *Home Folks*, 1: 117–27.

69. Bixby-Smith, *Adobe Days*, p. 254.

70. William Dean Howells, "A Boy's Town," in Henry Steele Commager, ed., *Selected Writings of William Dean Howells* (New York: Random House, 1950), p. 799.

71. Another potentially dangerous fire game did not even use real fire. "Fire!" was essentially an initiation ritual for newcomers. The new boy was given the role of fireman and was sent off in search of a fire. When he "found" one, he was to call out "Fire! Fire! Fire!" The other boys would then storm out of a makeshift "engine house" and spray the novice with a shower of stones. The game is described in *American Folk-Lore* 4 (1891); reference from Mitford M. Mathews, *Dictionary of Americanisms on Historical Principles* (Chicago: University of Chicago Press, 1951), p. 607.

72. Carl Bridenbaugh, *Cities in the Wilderness: The First Century of Urban Life in America* (New York: Alfred A. Knopf, 1960), p. 366.

73. George J. Richman, *History of Hancock County, Indiana: Its People, Industries, and Institutions* (Greenfield, Ind.: William Mitchell Printing Company, 1916), p. 603.

74. Quoted in Monica Kiefer, *American Children through Their Books, 1700–1835* (Philadelphia: University of Pennsylvania Press, 1948), p. 167.

75. Howells, "A Boy's Town," p. 729.

76. Brooks McNamara, *Step Right Up* (Garden City, N.Y.: Doubleday and Company, 1976), pp. 14–15.

77. John W. Reps, *Cities on Stone: Nineteenth-Century Lithograph Images of the Urban West* (Fort Worth: Amon Carter Museum, 1976), p. 34. See also Jadviga M. da Costa Nunes, "The Industrial Landscape in America, 1800–1840: Ideology into Art," *Industrial Archeology* 12, no. 2 (1986): 19–38.

78. The physical evidence of volcanoes made it easy to believe in subterranean hellfires of a very real sort. The word *brimstone* means "burning stone" but refers specifically to sulphur, which burns with a colorful flame and notoriously unpleasant odor.

79. Quoted in Charles Angoff, *A Literary History of the American People*, 2 vols. in 1 (New York: Tudor Publishing Company, 1935), 1: 133.

80. Quoted in David E. Stannard, *The Puritan Way of Death: A Study in Religion, Culture, and Social Change* (New York: Oxford University Press, 1977), p. 68.

81. Jonathan Edwards, "Sinners in the Hands of an Angry God," in *Jonathan Edwards: Representative Selections, with Introduction, Bibliography, and Notes*, ed. Clarence H. Faust and Thomas H. Johnson (New York: Hill and Wang, 1962), p. 165.

82. Quoted in Alice Felt Tyler, *Freedom's Ferment: Phases of American Social History from the Colonial Period to the Outbreak of the Civil War* (New York: Harper and Row, 1962), pp. 330–31.

83. Excerpt from John Ruskin, "Of Queen's Gardens," *Sesame and Lilies* (1864), quoted in Gwendolyn Wright, *Moralism and the Model Home: Domestic Architecture and Cultural Conflict in Chicago, 1873–1913* (Chicago: University of Chicago Press, 1980), p. 13.

84. An entire book could be written on the relationship of the hearth to the home. Many of the modern studies on housework cover this topic. See, for instance, Ruth Schwartz Cowan, *More Work for Mother: The Ironies of Household Technology from the Open Hearth to the Microwave* (New York: Basic Books, 1983), and Harvey Green, *The Light of the Home: An Intimate View of the Lives of Women in Victorian America* (New York: Pantheon Books, 1983). The minister referred to is Horace Bushnell, whose views on Christian nurturing are described in David P. Handlin, *The American Home: Architecture and Society, 1815–1915* (Boston: Little, Brown and Company, 1979), pp. 6–12. It is Handlin's own description of Harriet B. Stowe's "The Ravages of a Carpet"

that is quoted. Her story is a classic description of the hearth-warmed home in operation. See Handlin, p. 426.

85. Quoted in Elisabeth Donaghy Garrett, *At Home: The American Family 1750–1870* (New York: Harry N. Abrams, 1990), p. 140.

86. Jonathan Dawson, "Coming of the Pioneer," in Miller, ed., *Home Folks*, 1: 8–13.

87. "The Mechanics' Saturday Night," *Scientific American* 1, no. 12 (December 4, 1845).

88. Harriet Beecher Stowe, *Oldtown Folks* (Cambridge, Mass.: Harvard University Press, 1966), 105–8; Nathaniel Hawthorne, *The House of the Seven Gables, A Romance* (New York: Books, Inc., n.d.), p. 103.

89. One of the better known of these efforts is Anna Bache's *Scenes at Home: Or the Adventures of a Fire Screen* (Philadelphia, 1852). But even in more conventional novels, such as Louisa May Alcott's *Little Women*, the hearth tools are occasionally referred to as if they have lives of their own.

90. Quoted in Priscilla Joan Brewer, "Home Fires: Cultural Responses to the Introduction of the Cookstove, 1815–1900" (Ph.D. diss., Brown University, 1987), p. 16.

91. Lucy Larcom, *A New England Girlhood, Outlined from Memory* (1889; rpt. Williamstown, Mass.: Corner House Publishers, 1977), p. 23.

92. James Thomas Flexner, *History of American Painting*, vol. 2: *The Light of Distant Skies* (New York: Dover, 1969), p. 124.

93. Virginia Wilcox Ivins, *Pen Pictures of Early Western Days* (1905), excerpted in Fischer, ed., *Let Them Speak for Themselves*, pp. 75–82; *Andy Adams' Campfire Tales*, ed. Wilson M. Hudson (Lincoln and London: University of Nebraska Press, 1956), pp. 3–4.

94. Samuel G. Goodrich, *Enterprise, Industry and Art of Man, As Displayed in Fishing, Hunting, Commerce, Navigation, Mining, Agriculture and Manufactures* (Boston: Rand and Mann, 1849), p. iii.

95. Ellen H. Rollins, *New England Bygones, by E. H. Arr* (Philadelphia, 1880), pp. 37–38, quoted in Garrett, *At Home*, p. 108.

96. *The Poems of Henry Wadsworth Longfellow*, pp. 635–36.

97. Larcom, *New England Girlhood*, p. 23.

CHAPTER 3
BAD MASTER

1. One such tragedy was only narrowly averted in Boston on March 6, 1775, when a crowd, gathered to hear Joseph Warren speak at the Old South Meeting in commemoration of the Boston Massacre, mistook the words "Fie! Fie!" for "Fire!" Described in Esmond Wright, ed., *The Fire of Liberty* (New York: St. Martin's Press, 1983), pp. 17–18.

2. Mary C. Henderson, *Theater in America: 200 Years of Plays, Players, and Productions* (New York: Harry N. Abrams, 1986), pp. 243–44; David D. Dana, *The Fireman: The Fire Departments of the United States, With a Full Account of All Large Fires, Statistics of Losses and Expenses, Theatres Destroyed by Fire, and Accidents, Anecdotes, and Incidents* (Boston: James French and Company, 1858), pp. 251–52.

3. Quoted in Richard C. Wade, *The Urban Frontier: Pioneer Life in Early Pittsburgh,*

Cincinnati, Lexington, Louisville, and St. Louis (Chicago: University of Chicago Press, 1959), p. 91.

4. Stephen J. Pyne, *Fire in America: A Cultural History of Wildland and Rural Fire* (Princeton: Princeton University Press, 1982), p. xvi.

5. Frederick Overman, *The Manufacture of Iron, in All Its Various Branches* (Philadelphia: H. C. Baird, 1850), pp. 106–7.

6. Don H. Berkebile, "Wooden Roads," in Brooke Hindle, ed., *Material Culture of the Wooden Age* (Tarrytown, N.Y.: Sleepy Hollow Press, 1981), pp. 129–58.

7. William Allen White, *The Autobiography of William Allen White* (New York: Macmillan, 1946), p. 38.

8. "Burning Coal Mines," *Franklin Institute Journal* n.s., 26 (1840): 68–69; M. Darlington, "Mode of Extinguishing Fires in Coal Mines," *Franklin Institute Journal* 3d ser., 18 (1849): 152–54. Many other periodicals, including *Scientific American* and *Niles' Weekly Register*, offered contributions to this genre.

9. Christine Meisner Rosen, *The Limits of Power: Great Fires and the Process of City Growth in America* (Cambridge: Cambridge University Press, 1986), p. 178.

10. Quoted in Dana, *The Fireman*, p. 163.

11. John W. St. Croix, *Pictorial History of the Town of Hartford, Vermont, 1761–1963* (Orford, N.H.: Equity Publishing, 1963), pp. 103–4.

12. *Phillips' United States Patent Fire Annihilator* (New York: Office of the Company, 1851), p. 4.

13. Edith White, "Memories of Pioneer Childhood and Youth in French Corral and North San Juan, Nevada County, California," excerpted in Christiane Fischer, ed., *Let Them Speak for Themselves: Women in the American West, 1849–1900* (New York: Archon Books, 1977), pp. 271–82 (see p. 279).

14. Luzena Stanley Wilson, *Luzena Stanley Wilson, '49er: Memories Recalled Years Later for her Daughter, Correnah Wilson Wright* (Mills College, Calif.: Eucalyptus Press, 1937), excerpted in Fischer, ed., *Let Them Speak for Themselves*, pp. 151–65 (see pp. 163–64).

15. *Phillips' . . . Fire Annihilator*, p. 3.

16. Dana, *The Fireman*, p. 65.

17. A. L. Todd, *A Spark Lighted in Portland: The Record of the National Board of Fire Underwriters* (New York: McGraw-Hill, 1966), pp. 33–48.

18. Frederick Marryat, *A Diary in America: With Remarks on Its Institutions*, ed. Sydney Jackman (Westport, Conn.: Greenwood Press, 1962), p. 37; *Phillips' . . . Fire Annihilator*, p. 3.

19. Basil Hall, *Travels in North America* (1829; rpt. Graz: Akademische Druck- und Verlagsanstalt, 1965), 1: 19.

20. Dana, *The Fireman*, pp. 358–59.

21. Ibid., p. 365–66.

22. Todd, *A Spark Lighted in Portland*, pp. 10, 34–35.

23. Ibid., p. 42.

24. Statistics for these great fires are from Rosen, *The Limits of Power*, p. 92.

25. Quoted in John Carey, ed., *Eyewitness to History* (Cambridge, Mass.: Harvard University Press, 1987), pp. 418, 419.

26. A useful overview of this historic fire is found in Gordon Thomas and Max Morgan Witts, *The San Francisco Earthquake* (New York: Stein and Day, 1971).

27. Mrs. Laurence A. Curtis, letter of April 19, 1906, quoted in *The Washington Post*, October 10, 1989.

28. Clarence A. Glasrud, *Roy Johnson's Red River Valley: A Selection of Historical Articles First Printed in "The Forum" from 1941 to 1962* (Moorhead, Minn.: Red River Valley Historical Society, 1982), p. 242.

29. Quoted in *Reader's Digest Story of the Great American West* (Pleasantville, N.Y.: Reader's Digest Association, 1977), p. 315.

30. Pyne, *Fire in America*, pp. 204, 56, 199–200, 103. Pyne's study of this vast subject examines wildfire broadly within the larger social context.

31. Lee H. Nelson, "The Colossus of Philadelphia," in Hindle, ed., *Material Culture of the Wooden Age*, pp. 159–83.

32. Interestingly, historic houses that burned in place have the desirable attributes of high "focus" and "visibility," both of which aid the archaeologist in reconstructing architectural practices of the past. See James Deetz, *In Small Things Forgotten: The Archaeology of Early American Life* (New York: Doubleday, 1977), p. 95.

33. Mitchell Wilson, *American Science and Invention: A Pictorial History* (New York: Simon and Schuster, 1954), pp. 142–43.

34. Anne Bradstreet, *The Works of Anne Bradstreet*, ed. Jeannine Hensley (Cambridge, Mass.: Belknap Press of Harvard University Press, 1967), pp. 292–93.

35. Nina Churchman Larowe, "An Account of My Life's Journey so far: its Prosperity, its Adversity, its Sunshine and its Clouds," in Fischer, ed., *Let Them Speak for Themselves*, pp. 207–28 (see pp. 225–26).

36. Thomas P. Jones, *New Conversations on Chemistry, Adapted to the Present State of That Science . . . On the Foundation of Mrs. Marcet's "Conversations on Chemistry"* (Philadelphia: John Grigg, 1832), p. 110.

37. From A.F.M. Willick's *The Domestic Encyclopedia* (Philadelphia, 1804), quoted in Arthur H. Hayward, *Colonial and Early American Lighting* (1927; rpt. New York: Dover Publications, 1962), p. 84.

38. Edith Stratton Kitt, *Pioneering in Arizona: The Reminiscences of Emerson Oliver Stratton and Edith Stratton Kitt* (Tucson: Arizona Pioneers' Historical Society, 1964), excerpted in Fischer, ed., *Let Them Speak for Themselves*, pp. 283–97 (see p. 293).

39. Lillian Schlissel, *Women's Diaries of the Westward Journey* (New York: Schocken Books, 1982), p. 80.

40. "Lyman's Apparatus for Warming and Ventilating Rooms," *Scientific American* 9, no. 26 (March 11, 1854).

41. Quoted in Henry J. Kauffman, *The American Fireplace: Chimneys, Mantelpieces, Fireplaces & Accessories* (Nashville and New York: Thomas Nelson, 1972), p. 226.

42. In Schlissel, *Women's Diaries of the Westward Journey*, p. 83.

43. Quoted in Kauffman, *The American Fireplace*, p. 249.

44. Quoted in Elisabeth Donaghy Garrett, *At Home: The American Family 1750–1870* (New York: Harry N. Abrams, 1990), pp. 18–19.

45. Henry David Thoreau, *Walden and Other Writings*, ed. Brooks Atkinson (New York: Modern Library, 1950), p. 106.

46. "Apparatus for Condensing Smoke," *Scientific American* 9, no. 17 (January 7, 1854).

47. James Bryce, *The American Commonwealth*, 2 vols. (New York: Macmillan, 1912), 2: 805.

48. "Come To Blows!" *Scientific American* 1, no. 2 (September 4, 1845).

49. David Thomas, *Travels Through the Western Country in the Summer of 1816* (1819; rpt. Darien, Conn.: Hafner, 1970), p. 50.

50. Willard Glazier, *Peculiarities of American Cities* (1885), quoted in Mike Edelhart and James Tinen, *America the Quotable* (New York: Facts On File Publications, 1983), p. 391.

51. Duane A. Smith, *Mining America: The Industry and the Environment, 1800–1980* (Lawrence: University Press of Kansas, 1987), describes many problem areas, including Black Hawk, Colo. (p. 12), Leadville, Colo. (p. 11), and Butte, Mont. (p. 76).

52. James Walter Goldthwait, *Diary of Western Trip, 1902* (Burnsville, N.C.: Ad Lib Press, 1982), pp. 5, 21, 24, 26.

53. Jacob Bigelow, "Observations and Experiments on the Treatment of Injuries Occasioned by Fire and Heated Substances," *New England Journal of Medicine and Surgery* 1, no. 1 (1812): 52–64.

54. John C. Gunn, *Gunn's Domestic Medicine, or Poor Man's Friend, in the Hours of Affliction, Pain and Sickness*, 4th ed. (Madisonville: Edwards and Henderson, 1834), p. 311.

55. Daniel Drake, "History of Two Cases of Burn, Producing Serious Constitutional Irritation," *Western Journal of the Medical and Physical Sciences* 4, no. 1 (1830–1831): 48–60.

56. Gunn, *Gunn's Domestic Medicine*, pp. 311–12.

57. The Johnstown Flood claimed a total of 2,209 lives. For a good survey of this disaster, see Donald Dale Jackson, "When 20 Million Tons of Water Flooded Johnstown," *Smithsonian* 20 (May 1989): 50–61.

58. Interestingly, fatalities were not as numerous as might be expected from these huge fires. There were no deaths officially ascribed to the 1904 Baltimore fire and very few, if any, that could be traced directly to Portland's 1866 blaze or Boston's 1872 disaster, in part because these fires occurred primarily in business districts at times when people were not at work. At other times, there was often enough warning so that people could escape. The real problems were the fires that raged quickly through rooms, buildings, or areas with lots of people.

59. A useful survey is contained in Woody Gelman and Barbara Jackson, *Disaster Illustrated: Two Hundred Years of American Misfortune* (New York: Harmony Books, 1976), pp. 168–69.

60. Lucius Beebe and Charles Clegg, *Hear the Train Blow: A Pictorial Epic of America in the Railroad Age* (New York: E. P. Dutton, 1952), p. [78]. When the steamboat *Moselle* exploded in 1838, there were more than 200 fatalities; and as many as 300 people died in the aftermath of the *Montreal* disaster of 1857. See Gelman and Jackson, *Disaster Illustrated*, pp. 72–73.

61. Quoted in Lewis O. Saum, "Death in the Popular Mind of Pre–Civil War America," in David E. Stannard, ed. *Death in America* (Philadelphia: University of Pennsylvania Press, 1975), pp. 30–48.

62. William H. Brown, quoted in Oliver Jensen, *The American Heritage History of Railroads in America* (New York: American Heritage Publishing Company, 1975), p. 25.

63. Quoted in John H. White Jr., *The American Railroad Passenger Car* (Baltimore: Johns Hopkins University Press, 1978), p. 393.

64. Harold Hancock, *The History of Nineteenth Century Laurel (Del.)* (Laurel: Laurel Historical Society, 1983), p. 318.

65. Quoted in Benjamin Rush, *The Autobiography of Benjamin Rush: His "Travels Through Life" Together with His Commonplace Book for 1789–1813*, ed. George W. Corner (Princeton: Princeton University Press, for the American Philosophical Society, 1948), p. 238.

66. Catharine E. Beecher and Harriet Beecher Stowe, *The American Woman's Home* (New York: J. B. Ford and Company, 1869), p. 352.

67. Alexander V. Hamilton, *The Household Cyclopaedia of Practical Receipts and Daily Wants* (Springfield: W. J. Holland, 1875), pp. 390–91.

68. Elizabeth F. Ellet, *The Pioneer Women of the West* (1852; rpt. Freeport, N.Y.: Books for Libraries Press, 1973), p. 134.

69. Lydia Maria Child, *The American Frugal Housewife, Dedicated to Those Who Are Not Ashamed of Economy*, 12th ed. (1833; rpt. Cambridge, Mass.: Applewood Books, n.d.), pp. 28–29.

70. Skin grafting was attempted as early as 1854, but progress in using the technique was slow over the next fifty years. The journal literature of the period refers intermittently to successful cases but concentrates much more heavily on discussions of ointments and dressings.

71. Page Smith, *Daughters of the Promised Land* (Boston: Little, Brown and Company, 1970), p. 55.

72. *The Cottage Physician, for Individual and Family Use* (Springfield, Mass.: King-Richardson Company, 1900), p. 458.

73. Hannah Barnard, "Reflections and Experiments on Burns, and Their Treatment," *Medical Repository* n.s., 5 (1820): 90–94.

74. There were an estimated 2,909 deaths from burns and scalds in England in 1845. One may assume that the American count was at least as high. See "Accident and Crime," *Scientific American* 1, no. 10 (November 20, 1845).

75. *A Tribute to Woody Guthrie*, words and music by Woody Guthrie, script by Millard Lampell (New York: Ludlow Music Inc., 1972), pp. 63–64.

76. Michael Lesy, *Wisconsin Death Trip* (New York: Pantheon, 1973), "1898."

77. William Byrd, *The Great American Gentleman: William Byrd of Westover in Virginia; His Secret Diary for the Years 1709–1712*, ed. Louis B. Wright and Marion Tinling (New York: Capricorn Books 1963), p. 89.

78. Mark Twain, *Personal Recollections of Joan of Arc, by the Sieur Louis de Conte, Translated by Jean François Alden* (New York: Harper and Row, 1924), pp. 281–82.

79. For example, Ellet, *Pioneer Women of the West*, p. 126, tells the story of a mother and daughter who were seized by Cherokees, tortured by fire, and finally burned at the stake over the course of three agonizing days.

80. These are the words of the Reverend John Wilson of Boston, quoted in John C. Miller, *The First Frontier: Life in Colonial America* (New York: Dell, 1966), p. 77.

81. John F. Watson, *Annals of Philadelphia and Pennsylvania*, 3 vols. (Philadelphia: J. M. Stoddart, 1879–1881), 1: 309.

82. John Hope Franklin, *From Slavery to Freedom: A History of Negro Americans*, 6th ed. (New York: Alfred A. Knopf, 1988), pp. 226 and passim.

83. Richard Hofstadter and Michael Wallace, *American Violence: A Documentary History* (New York: Alfred A. Knopf, 1970), p. 187.

84. Winthrop D. Jordan, *White Over Black: American Attitudes Toward the Negro, 1550–1812* (Chapel Hill: University of North Carolina Press, 1968), pp. 398 (1805 incident) and 392 (1806 incident).

85. Marryat, *A Diary in America*, p. 38. Given the uneasiness of race relations in New York at the time, it is not unlikely that "notorious carelessness" often encompassed "intentional laxity" with fire.

86. Ibid., p. 328.

87. Ernest Earnest, *The Volunteer Fire Company, Past and Present* (New York: Stein and Day, 1979), p. 92.

88. Dana, *The Fireman*, p. 64.

89. There are many descriptions of these riots; an excellent overview of the subject is contained in Hofstadter and Wallace, *American Violence*. The statistics for riots from 1830 to 1850 are Richard Maxwell Brown's, quoted in Hofstadter and Wallace, pp. 14–15.

90. Ibid., p. 343.

91. Page Smith, *Trial by Fire: A People's History of the Civil War and Reconstruction* (New York: McGraw-Hill, 1982), p. 382, refers to fire this way in terms of slave rebellion specifically, but it describes black/white relations generally as well.

92. Earnest, *The Volunteer Fire Company*, pp. 88–89.

93. Lawrence Friedman, *A History of American Law* (New York: Simon and Schuster, 1973), p. 200; Jordan, *White Over Black*, pp. 392–93, 404; *Encyclopedia of Crime and Justice*, ed. Sanford H. Kadish (New York: Free Press, 1983), 1: 133.

94. Alfred M. Downes, *Fire Fighters and Their Pets* (New York and London: Harper and Brothers, 1907), pp. 84–85.

95. *Report of the Committee on Incendiarism and Arson, Made to the National Board of Fire Underwriters, May 19, 1881* (New York: Styles and Cash, 1881), p. 27.

96. Carl Bridenbaugh, *Cities in the Wilderness: The First Century of Urban Life in America* (New York: Alfred A. Knopf, 1960), p. 368.

97. *Report . . . on Incendiarism and Arson*, p. 7.

98. "Very Alarming to Incendiaries, &c.," *Scientific American* 1, no. 17 (January 8, 1846).

99. *Report . . . on Incendiarism and Arson*, pp. 8, 14–16.

100. Hofstadter and Wallace, *American Violence*, pp. 211–13.

101. James R. Chiles, "County Seats Were a Burning Issue in the Wild West," *Smithsonian* 20, no. 12 (March 1990): 100–110.

102. Pennsylvania and Connecticut men: *Report . . . on Incendiarism and Arson*, pp. 8–12; Wisconsin man: Lesy, *Wisconsin Death Trip*, "1897"; *Rochester Herald*: *Report . . . on Incendiarism and Arson*, p. 30.

103. Lesy, *Wisconsin Death Trip*, "1898" and "1900."

104. Norse mythology underscores this point by portraying arson as unfair fighting.

105. *Report . . . on Incendiarism and Arson*, p. 16.

106. Wilson, *Luzena Stanley Wilson, '49er*, in Fischer, ed., *Let Them Speak for Themselves*, pp. 151–65 (see pp. 159–60).

107. Quoted in Miller, *The First Frontier*, p. 175.

108. Ellen McGowan Biddle, *Reminiscences of a Soldier's Wife* (Philadelphia: J. B. Lippincott Company, 1907), excerpted in Fischer, ed., *Let Them Speak for Themselves*, pp. 124–36.

109. Glenda Riley, *Frontierswomen: The Iowa Experience* (Ames: Iowa State University, 1981), p. 44.

110. Herman Melville, *White-Jacket, or The World in a Man-of-War*, with introduction by Alfred Kazin (New York: New American Library/Signet, 1979), p. 88.

111. Bridenbaugh, *Cities in the Wilderness*, pp. 11–12, 151–52.

112. "Jack Frost," *Scientific American* 1, no. 13 (December 11, 1845).

113. Margaret M. Coffin, *Death in Early America: The History and Folklore of Customs and Superstitions of Early Medicine, Funerals, Burials, and Mourning* (New York: Thomas Nelson, 1976), p. 181.

CHAPTER 4

FIGHTING BACK

1. *Phillips' United States Patent Fire Annihilator* (New York: Office of the Company, 1851), pp. 3–4.

2. "Fires in our City—The Annihilator," *Scientific American* 9, no. 19 (January 21, 1854).

3. Carl Bridenbaugh, *Cities in the Wilderness: The First Century of Urban Life in America* (New York: Alfred A. Knopf, 1960), p. 56.

4. Christine Meisner Rosen, *The Limits of Power: Great Fires and the Process of City Growth in America* (Cambridge: Cambridge University Press, 1986), pp. 212–18.

5. David D. Dana, *The Fireman: The Fire Departments of the United States, With a Full Account of All Large Fires, Statistics of Losses and Expenses, Theatres Destroyed by Fire, and Accidents, Anecdotes, and Incidents* (Boston: James French and Company, 1858), p. 147.

6. Russell F. Weigley, ed., *Philadelphia: A 300-Year History* (New York: W. W. Norton and Company, 1982), p. 316; George J. Richman, *History of Hancock County, Indiana: Its People, Industries, and Institutions* (Greenfield, Ind.: William Mitchell Printing Company, 1916), p. 603; Richard C. Wade, *The Urban Frontier: Pioneer Life in Early Pittsburgh, Cincinnati, Lexington, Louisville, and St. Louis* (Chicago: University of Chicago Press, 1959), pp. 92 (Pittsburgh), 90–91 (Cincinnati).

7. [Note], *Scientific American* 1, no. 18 (January 15, 1846).

8. John S. Duss, *The Harmonists: A Personal History* (1943; rpt. Ambridge, Pa.: The Harmonie Associates, 1970), pp. 191–92.

9. Alexander V. Hamilton, *The Household Cyclopaedia of Practical Receipts and Daily Wants* (Springfield: W. J. Holland, 1875), pp. 390–91.

10. Catharine E. Beecher and Harriet Beecher Stowe, *The American Woman's Home* (New York: J. B. Ford and Company, 1869), pp. 360–61; Hamilton, *Household Cyclopaedia*, p. 390.

11. Denison Olmsted, *Observations on the Use of Anthracite Coal* (Cambridge, Mass.: Folsom, Wells, and Thurston, 1836), p. 11; John Lee Comstock, *Elements of Chemistry* . . . (New York: Pratt, Woodford and Company, 1853), p. 380; "Another Gas Explosion," *Scientific American* 9, no. 24 (February 25, 1854).

12. Hamilton, *Household Cyclopaedia*, p. 20.

13. *Asbestos* is a descriptive mineralogical term for a variety of minerals that adopt a fibrous habit. When woven or matted, these fibers provide excellent heat insulation. Some of these minerals, such as the chrysotile form of serpentine, possess a layered atomic structure that rolls up like microscopic carpets. Other "asbestoform" minerals, such as the pyroxenes, have chainlike atomic structures that contribute to the needlelike grains. Confusion has arisen because only a few of these types of asbestos pose a serious health risk, yet all are banned by current legislation. For a current review, see H.C.W. Skinner, M. Ross, and C. Frondel, *Asbestos and Other Fibrous Minerals: Mineralogy, Crystal Chemistry, and Health Effects* (New York: Oxford University Press, 1988).

14. "Severe and Destructive Lightning," *Scientific American* 1, no. 2 (September 4, 1845).

15. Sears, Roebuck and Company, *The 1902 Edition of the Sears Roebuck Catalogue* (New York: Crown Publishers, 1969), p. 567.

16. Quoted in Dana, *The Fireman*, p. 155.

17. Ibid., p. 135.

18. E. D. Bryan, "We Are at the Mercy of the Flames: The Great Laurel Fire," in Harold Hancock, *The History of Nineteenth Century Laurel (Del.)* (Laurel: Laurel Historical Society, 1983), pp. 326–43 (see pp. 335–36).

19. Ibid., pp. 326–43.

20. Numerous fascinating accounts of firefighting have appeared over the years, including Herbert Asbury, *Ye Olde Fire Laddies* (New York: Alfred A. Knopf, 1930); Walter P. McCall, *American Fire Engines Since 1900* (Glen Ellyn, Ill.: Crestline Publishers, 1976); Donald J. Cannon, *Heritage of Flames* (New York: Doubleday, 1977); Dennis Smith, *Dennis Smith's History of Firefighting in America: 300 Years of Courage* (New York: Dial Press, 1978); Ernest Earnest, *The Volunteer Fire Company, Past and Present* (New York: Stein and Day, 1979). Local histories furnish innumerable stories, and professional journals such as *Fire Journal* and *Fire and Water* help fill in details.

21. Bridenbaugh, *Cities in the Wilderness*, p. 58.

22. "Improved Fire Engine," *Scientific American* 1, no. 6 (October 2, 1845).

23. Earnest, *The Volunteer Fire Company*, p. 109.

24. Quoted in ibid., p. 9.

25. Rebecca Zurier, *The American Firehouse: An Architectural and Social History* (New York: Abbeville Press, 1982), p. 74; Weigley, *Philadelphia*, pp. 346–48.

26. Quoted in Dana, *The Fireman*, pp. 244–48.

27. Philadelphia installed piped water in 1801, but only with the establishment of the Fairmount Waterworks in 1815 was there enough pressure for firefighting. New York City's famous Croton Reservoir opened in 1842; Boston got its system in 1848.

28. William Dean Howells, "A Boy's Town," in Henry Steele Commager, ed., *Selected Writings of William Dean Howells* (New York: Random House, 1950), p. 861.

29. Clarence A. Glasrud, *Roy Johnson's Red River Valley: A Selection of Historical Articles First Printed in "The Forum" from 1941 to 1962* (Moorhead, Minn.: Red River

Valley Historical Society, 1982), p. 326. Volunteer firefighters remain an important element in local fire programs. Ernest Earnest (*The Volunteer Fire Company*, p. 2) estimates that as of 1979 there were more than 24,000 volunteer and "part-paid" fire departments in the United States. The foregoing analysis in no way reflects on the exemplary performances of present-day volunteers.

30. The word *buff* refers to the buff-colored overcoat worn by the volunteer firemen in New York City circa 1820, but the word has become generalized to mean any enthusiast about firefighting.

31. William Allen White, *The Autobiography of William Allen White* (New York: Macmillan, 1946), p. 38.

32. Beecher and Stowe, *American Woman's Home*, pp. 360–61.

33. Hamilton, *Household Cyclopaedia*, p. 390.

34. Aetna Company advertisement, published in *Michigan State Gazetteer* (Detroit, 1863), facing p. 508.

35. Hamilton, *Household Cyclopaedia*, p. 209; *The Universal Self-Instructor and Manual of General Reference* (1883; rpt. New York: Winter House, 1970), p. 179; *The American Domestic Cyclopedia, a Volume of Universal Ready Reference for American Women in American Homes* (New York: F. M. Lupton, 1890), p. 531.

36. Many books cover the development of the insurance industry in America. A good summary appears in Irving Pfeffer and David R. Klock, *Perspectives on Insurance* (Englewood Cliffs, N.J.: Prentice-Hall, 1974), who state that the San Francisco disaster bankrupted many companies (p. 47). A. L. Todd claims that only twenty major insurers went under after that disaster in contrast to the higher failure rate quoted for earlier fires (Todd, *A Spark Lighted in Portland: The Record of the National Board of Fire Underwriters* [New York: McGraw-Hill, 1966], p. 46). See also John Bainbridge, *Biography of an Idea: The Story of Mutual Fire and Casualty Insurance* (Garden City, N.Y.: Doubleday, 1952), and Andrew Tobias, *The Invisible Bankers: Everything the Insurance Industry Never Wanted You to Know* (New York: Linden Press, 1982).

37. "Aldridge Notes," *The Livingston Enterprise* (January 25, 1902), quoted in Bill and Doris Whithorn, *A Photo History of Aldridge, Coal Camp That Died A-Bornin'* (Minneapolis: Acme Printing, 1966), p. 163.

38. A good summary of this story appears in *Reader's Digest Story of the Great American West* (Pleasantville, N.Y.: Reader's Digest Association, 1977), p. 353.

39. Quoted in Zurier, *The American Firehouse*, p. 15.

40. The extraordinary range of artifacts relating to the firefighting fraternity is described in some detail in Chuck Deluca, *Firehouse Memorabilia: A Collector's Reference* (York, Maine: Maritime Antique Auctions, 1989).

41. The dalmatian has come to epitomize the firehouse pet. In fact, because of their intelligence, dalmatians often served as fire dogs. They not only kept firemen company but also barked loudly to help clear the streets when the fire engines rolled and protected the firehouse when the men were gone. Sometimes, they went to fires as well. (Good dogs were awarded special insignias to wear on their collars.) Not all fire pets were dalmatians, however, as Alfred M. Downes points out in *Fire Fighters and Their Pets* (New York: Harper and Brothers, 1907). In New York City, other breeds (such as the St. Bernard) were known to serve. Cats and even monkeys were also kept as firehouse pets.

42. Quoted in Donald L. Collins, *Firefighting in Olde Lancaster* (New Holland, Pa.: Creative Communications Group, 1976), p. 42.

43. Quoted in John Wilmerding, *Important Information Inside: The Art of John F. Peto and the Idea of Still-Life Painting in Nineteenth-Century America* (Washington, D.C.: National Gallery of Art, 1983), p. 33.

44. Geoffrey N. Stein, *From the Collections: The Fire Apparatus at the New York State Museum* (Albany: University of the State of New York, 1987), pp. 10–11.

45. William Dean Howells, "A Boy's Town," p. 861.

46. Quoted in Collins, *Firefighting in Olde Lancaster*, p. 42.

47. Downes, *Fire Fighters and Their Pets*, p. 11.

48. The firm of Currier and Ives had particular interest in the subject since both proprietors, as well as a number of their employees, were volunteer firemen. The company issued two sets of prints with firemen as the theme: "The American Fireman" series (in which the view entitled "Always Ready" was a portrait of Nathaniel Currier) and "The Life of a Fireman" series (which consisted of four sequential views of fighting a fire, to which were added two additional prints several years later).

49. When in 1872 Gilmore organized a follow-up performance, the World's Peace Jubilee, he could not resist inviting the firemen again.

50. A good discussion of Mose appears in Richard M. Dorson, *America in Legend: Folklore from the Colonial Period to the Present* (New York: Pantheon, 1973), pp. 99–108. See also Dorson, *American Folklore* (Chicago: University of Chicago Press, 1959), and Dorson, *Handbook of American Folklore* (Bloomington: Indiana University Press, 1983).

51. These words formed the subtitle of Increase Mather's sermon "Burnings Bewailed," delivered after the Boston fire of October 2, 1711. Cotton Mather and other Puritan divines preached on similar topics.

52. James Fenimore Cooper, *Home as Found* (New York: Capricorn Books, 1961), p. 107.

53. Nathaniel Hawthorne, *The House of the Seven Gables, A Romance* (New York: Books, Inc., n.d.), p. 148.

54. Quoted in Mike Edelhart and James Tinen, *America the Quotable* (New York: Facts On File Publications, 1983), p. 133.

55. Quoted in Paul B. Streng, "The Fire Department," in *The Howell Bicentennial History* (Howell, Mich.: American Revolution Bicentennial Committee, 1975), p. 347.

56. Lyon and Healy, *Band Instruments, Uniforms, Trimmings, &c.* (Chicago, 1881), p. 186.

57. Phineas T. Barnum, *Struggles and Triumphs: or Forty Years' Recollections of P. T. Barnum* (Hartford: J. B. Burr and Company, 1869), p. 702.

58. Kenneth E. Boulding, *The Image: Knowledge in Life and Society* (Ann Arbor: University of Michigan Press, 1956).

59. Quoted in Arthur D. Pierce, *Iron in the Pines: The Story of New Jersey's Ghost Towns and Bog Iron* (New Brunswick, N.J.: Rutgers University Press, 1957), p. 112.

60. Quoted in Longin Pastusiak, "Henryk Sienkiewicz, 1846–1916," in Marc Pachter, ed., *Abroad in America: Visitors to the New Nation, 1776–1914* (Reading, Mass.: Addison-Wesley with the National Portrait Gallery, 1976), pp. 176–85 (see p. 181).

CHAPTER 5
KEEP THE HOME FIRES BURNING

1. Susan Warner, *The Wide Wide World* (1851; rpt. New York: Feminist Press, 1987), pp. 10, 13. This enormously popular book was the first American novel to sell a million copies.

2. Mary Bennett diary in Cornell University Archives, quoted in Ruth Schwartz Cowan, *More Work for Mother: The Ironies of Household Technology from the Open Hearth to the Microwave* (New York: Basic Books, 1983), pp. 61–62; Jonathan Dawson, "Coming of the Pioneer," in Marguerite Miller, ed., *Home Folks* (Marceline, Mo.: Walsworth, n.d.), 1: 8–13; Helen M. Carpenter, quoted in Lillian Schlissel, *Women's Diaries of the Westward Journey* (New York: Schocken Books, 1982), p. 78.

3. Deborah J. Warner, "The Women's Pavilion," in Robert C. Post, ed., *1876: A Centennial Exhibition* (Washington, D.C.: Smithsonian Institution, 1976), pp. 162–73.

4. Frank Rowsome Jr., *The Bright and Glowing Place* (Brattleboro, Vt.: Stephen Greene Press, 1975), p. 77.

5. The expression "borrowing fire" has a long history relating to the widespread belief that stolen fire is tainted with wickedness. To this day we ask to "borrow a light" in recognition of the ancient taboo against taking fire from others.

6. David W. Shryock, "Incidents of Boyhood Days," in Miller, ed., *Home Folks*, 2: 49–56 (see p. 51).

7. Edward Everett Hale, *A New England Boyhood* (1893; rpt. Upper Saddle River, N.J.: Literature House, 1970), p. 118.

8. Documents relating to these and many other match producers and distributors are available in the Warshaw Collection of Business Americana at the Archives Center, National Museum of American History, Washington, D.C.

9. Shryock, "Incidents of Boyhood Days," p. 51.

10. For a history of the company and its activities, see Herbert Manchester, *The Diamond Match Company: A Century of Service, of Progress, and of Growth, 1835–1935* (New York and Chicago: Diamond Match Company, 1935).

11. See, for example, Walter Hough, *Fire as an Agent in Human Culture* (Washington, D.C.: Government Printing Office, 1926), p. 53.

12. William Cronon, *Changes in the Land: Indians, Colonists, and the Ecology of New England* (New York: Hill and Wang, 1983), pp. 120–21.

13. Catharine E. Beecher and Harriet Beecher Stowe, *The American Woman's Home* (New York: J. B. Ford and Company, 1869), p. 360.

14. *The American Domestic Cyclopedia, a Volume of Universal Ready Reference for American Women in American Homes* (New York: F. M. Lupton, 1890), p. 531.

15. John H. White Jr., "Railroads: Wood to Burn," in Brooke Hindle, ed., *Material Culture of the Wooden Age* (Tarrytown, N.Y.: Sleepy Hollow Press, 1981), pp. 184–224 (see p. 199).

16. Ruth Schwartz Cowan has discussed this change in some detail in *More Work for Mother*. It applies to other sorts of housework as well.

17. Harriet Beecher Stowe, *Oldtown Folks* (Cambridge, Mass.: Harvard University Press, 1966), 104–5.

18. Susan Strasser, *Never Done: A History of American Housework* (New York: Pantheon, 1982), p. 41.

19. Quoted in Faye E. Dudden, *Serving Women: Household Service in Nineteenth-Century America* (Middletown, Conn.: Wesleyan University Press, 1983), p. 131.

20. A.F.M. Willich, *The Domestic Encyclopedia; or, A Dictionary of Facts and Useful Knowledge*, 2d ed. (Philadelphia, 1821), 1: 373.

21. *Michigan State Gazetteer* (Detroit, 1863), pp. 464–65.

22. Margaret Hindle Hazen and Robert M. Hazen, *Wealth Inexhaustible: A History of America's Mineral Industries to 1850* (New York: Van Nostrand Reinhold, 1985), p. 202.

23. Lura Woodside Watkins, *American Glass and Glassmaking* (New York: Chanticleer Press, 1950), p. 10.

24. Lydia Maria Child, *The American Frugal Housewife, Dedicated to Those Who Are Not Ashamed of Economy*, 12th ed. (1833; rpt. Cambridge, Mass.: Applewood Books, n.d.), pp. 18–19, 11–12.

25. George Appleton, 1850, quoted in George Lewis Phillips, *American Chimney Sweeps: An Historical Account of a Once Important Trade* (Trenton, N.J.: Past Times Press, 1957), p. 54. Phillips demonstrates a clear relationship between the color of the sweepers' skin (and the fact that dark skins did not show the dirt) and the indifference of the American public toward abuses in the industry.

26. Denison Olmsted, *Observations on the Use of Anthracite Coal* (Cambridge, Mass.: Folsom, Wells, and Thurston, 1836), p. 10.

27. Rachel Haskell, "A Literate Woman in the Mines: The Diary of Rachel Haskell," in Christiane Fischer, ed., *Let Them Speak for Themselves: Women in the American West, 1849–1900* (New York: Archon Books, 1977), pp. 58–72 (see p. 67).

28. Device in authors' collection.

29. Sears, Roebuck and Company, *The 1902 Edition of the Sears Roebuck Catalogue* (New York: Crown Publishers, 1969), p. 578.

30. G.H.W. Bates & Company, *32d Annual Catalogue* (Boston: Bates & Company, ca. 1895), unpaged; see announcement for "Falls Heater."

31. A good overview of this subject appears in Jane C. Nylander, "Come, Gather Round the Chimney," *Natural History* 90 (1981): 98–103.

32. Rachel Haskell, "A Literate Woman in the Mines," pp. 58–72 (see p. 66).

33. Quoted in Elisabeth Donaghy Garrett, *At Home: The American Family 1750–1870* (New York: Harry N. Abrams, 1990), p. 149.

34. Although the use of such highly flammable materials for chimney construction was a dangerous practice, it persisted into the nineteenth century in rural areas.

35. Quoted in Russell Lynes, *The Domesticated Americans* (New York: Harper and Row, 1963), p. 25.

36. Child, *American Frugal Housewife*, p. 72.

37. Lucy Larcom, *A New England Girlhood, Outlined from Memory* (1889; rpt. Williamstown, Mass.: Corner House Publishers, 1977), p. 22.

38. George Wise, "Reckless Pioneer," *American Heritage of Invention and Technology* 6, no. 1 (Spring 1990): 26–31.

39. A good discussion of this evolution appears in Priscilla Joan Brewer, "Home

Fires: Cultural Responses to the Introduction of the Cookstove, 1815–1900" (Ph.D. diss., Brown University, 1987).

40. Loris S. Russell, "Early Nineteenth-Century Lighting," in Charles E. Peterson, ed., *Building Early America: Contributions Toward the History of a Great Industry* (Radnor, Pa.: Chilton Book Company, 1976), pp. 186–201.

41. William Chauncy Langdon describes wood this way, in contradistinction to coal, which enters vertically from the basement. See Langdon, *Everyday Things in American Life, 1776–1876*, 2 vols. (New York: Charles Scribner's Sons, 1955), 2: 198.

42. William Allen White includes a wonderful description of the "princely" woodshed of his youth in *The Autobiography of William Allen White* (New York: Macmillan, 1946), pp. 57–58. The use of the woodshed for punishment is, of course, legendary.

43. Special stoves were even developed to burn prairie grass, which was twisted into tight cylinders and inserted as needed.

44. *The Art of Cooking by Gas* (Published for the William M. Crane Company, 1898), p. 7.

45. Quoted in Schlissel, *Women's Diaries of the Westward Journey*, pp. 62–64.

46. William Dean Howells, "A Boy's Town" in Henry Steele Commager, ed., *Selected Writings of William Dean Howells* (New York: Random House, 1950), p. 775.

47. One derivation of the word *chauffeur* can be traced as follows. During the 1790s, a band of robbers roamed France under the leadership of one Jean l'Ecorcheur ("Jack the Scorcher"). He and his men enforced their demands for loot by tying their victims to a chair and pushing their feet into the hearth fire, for which outrages they became known as "chauffeurs," or firemen, from the French verb *chauffer*, to heat up or stoke. See Charles Earle Funk, *Thereby Hangs a Tale: Stories of Curious Word Origins* (New York: Harper and Row, 1985), p. 69.

48. 1660–1860 estimate: Cowan, *More Work for Mother*, pp. 28–29 and passim; 15–30 percent estimate: Dudden, *Serving Women*, p. 1.

49. Philip Vickers Fithian, *Journal & Letters of Philip Vickers Fithian, 1773–1774: A Plantation Tutor of the Old Dominion*, ed. Hunter Dickinson Farish (Williamsburg: Colonial Williamsburg, 1945), p. 53.

50. Sarah Josepha Hale, *The New Household Receipt-Book* (New York: H. Long and Brother, 1853), p. 252, as quoted in Dudden, *Serving Women*, p. 128.

51. Russell Lynes, *Domesticated Americans*, pp. 39–41.

52. Charles L. Eastlake cited this "absurdity of design" in his *Hints on Household Taste* (1878; rpt. Dover Publications, 1969), pp. 92–93.

53. Henry J. Kauffman, *The American Fireplace: Chimneys, Mantelpieces, Fireplaces & Accessories* (Nashville and New York: Thomas Nelson, 1972), p. 42.

54. Lina and Adelia B. Beard, *The American Girls Handy Book: How To Amuse Yourself and Others* (1887; rpt. Rutland, Vt.: Charles E. Tuttle Company, 1969), p. 451.

55. Quoted in Clifford Edward Clark Jr., *The American Family Home, 1800–1960* (Chapel Hill: University of North Carolina Press, 1986), p. 118.

56. National Radiator Company, *Beauty and Warmth: Aero Radiators and National Boilers* (Johnstown, Pa.: National Radiator Company, 1926), p. 3.

CHAPTER 6

UNDERSTANDING FIRE

1. See, for example, Stephen Toulmin and June Goodfield, *The Fabric of the Heavens* (New York: Harper and Row, 1961).

2. Joseph Black, *Lectures on the Elements of Chemistry, Delivered in the University of Edinburgh . . . Now Published from His Manuscripts by John Robison*, 2 vols. (Edinburgh: Mundell and Son for Longman and Rees, 1803); see also the first American edition (3 vols.; Philadelphia: M. Carey, 1806–1807). Black's caloric theory attempted to explain the increase in the weight of metal after oxidation and thus attributed mass to the caloric fluid. Other variants of the caloric theory, especially those proposed in the nineteenth century after oxidation was recognized as a chemical reaction, incorporated a weightless caloric fluid.

3. For an account of Benjamin Thompson's life and work see Sanborn C. Brown, *Benjamin Thompson, Count Rumford* (Cambridge, Mass.: MIT Press, 1979).

4. The original statement of thermodynamic principles appeared in Rudolf Julius Clausius, "Ueber die bewegende Kraft der Wärme, und die Gesetze, welche sich daraus für die Wärmelehre selbst albeiten lassen," *Annalen der Physik* 79 (1850): 368–97, 500–524. These ideas are presented in expanded form in an English translation, *The Mechanical Theory of Heat* (London: J. Van Voorst, 1867).

5. Confusion may arise over two different uses of the word *conserve*. The first law of thermodynamics, "energy is conserved," refers only to the fact that energy is neither created nor destroyed. Admonitions to "conserve energy" focus on the need to consume concentrated energy sources with restraint and thus relate to the second law of thermodynamics.

6. "Coal Trade," *Hazard's Register of Pennsylvania* 9 (1832): 79; 15 (1835): 141. See also, Margaret Hindle Hazen and Robert M. Hazen, *Wealth Inexhaustible: A History of America's Mineral Industries to 1850* (New York: Van Nostrand Reinhold, 1985), pp. 266–67.

7. A short, entertaining account of the history of temperature scales is provided by Henry Carrington Bolton, *Evolution of the Thermometer, 1592–1743* (Easton, Pa.: Chemical Publishing Company, 1900). A modern treatment with some historical review is given by T. J. Quinn, *Temperature* (New York: Academic Press, 1983). See also the epic series by the American Institute of Physics, *Temperature, Its Measurement and Control in Science and Industry* (New York: Reinhold Publishing, 1941–); vol. 4, pt. I, "Basic Methods" is particularly relevant.

8. Richard Morris, *Time's Arrow* (New York: Touchstone Books, 1985), pp. 109–35.

9. Reprints of classic papers by Boltzmann, Maxwell, and others appear in S. G. Brush, ed., *Kinetic Theory*, 2 vols. (Oxford: Oxford University Press, 1965–1966). Tyndall's contributions were collected in his treatise, *Heat Considered as a Mode of Motion* (London: Longman, Green, Longman, Roberts, and Green, 1863); see also the American edition (New York: D. Appleton, 1863).

10. Einstein's work on Brownian motion is described in George Gamow, *One, Two, Three . . . Infinity* (New York: Dover, 1974).

11. J. R. Partington, *A History of Chemistry*, 2 vols. (London: Macmillan, 1961),

2: 637–90. An English translation of some of Stahl's writings appears in Henry M. Leicester and Herbert S. Klickstein, eds., *A Source Book in Chemistry 1400–1900* (Cambridge, Mass.: Harvard University Press, 1968), pp. 58–63.

12. Joseph Priestley, *Experiments and Observations on Different Kinds of Airs, and Other Branches of Natural Philosophy*, 3 vols. (Birmingham: T. Pearson, 1790). Relevant portions of this work are reprinted in Leicester and Klickstein, eds., *Source Book in Chemistry*, pp. 112–23.

13. Lavoisier reached his widest audience with a general treatise on chemistry, *Traité élémentaire de Chimie* (Paris: Cuchet, 1789). A British edition appeared a year later (Edinburgh: W. Creech, 1790), but the first American edition waited considerably longer (Philadelphia: M. Carey, 1799). Important sections of this work are reprinted in Leicester and Klickstein, eds., *Source Book in Chemistry*, pp. 154–80.

14. Humphry Davy, *Elements of Chemical Philosophy* (London: J. Johnson, 1812); see also the first American edition (Philadelphia and New York: Bradford and Inskeep, 1812).

15. A general introduction to the physics and chemistry of the candle is presented by John W. Lyons, *Fire* (New York: Scientific American Books, 1985). A survey of modern fire research is given by Bruce Fellman, "Researchers Climb Inside of the Fire to Tweak the Flame," *Smithsonian* 18 (October 1987): 70–79.

16. James Clerk Maxwell's work is summarized in *A Treatise on Electricity and Magnetism* (Oxford: Clarendon Press, 1873).

17. Portions of this section are derived from Hazen and Hazen, *Wealth Inexhaustible*, pp. 197–203, 247–72.

18. Friedrich Knapp, *Chemical Technology; or, Chemistry Applied to the Arts and to Manufactures* (Philadelphia: Lea and Blanchard, 1848), pp. 22–24.

19. Wood pyrolysis is reviewed and colorfully illustrated by William H. Cottrell Jr., *The Book of Fire* (Missoula, Mont.: Mountain Press, 1989).

20. Detailed accounts of charcoaling are given by Frederick Overman, *The Manufacture of Iron, in All Its Various Branches* (Philadelphia: H. C. Baird, 1850), and Knapp, *Chemical Technology*.

21. Advocates for the inorganic origin of coal included William Byrd Powell, "Mineral Coal Is Not a Vegetable," *New-Yorker* 8 (1839): 2–3, 18–19, 33–34, 50–51, and Charles Whittlesey, "Bituminous Coal: Suggestions in Opposition to the Theory of the Vegetable Origin of Mineral Coal," *Hesperian* 1 (1838): 111–14, and *A Dissertation upon the Origin of Mineral Coal* (Cleveland: M. C. Younglove, 1846). The opposite point of view was argued by many geologists, including John Leonard Riddell, "On the Vegetable Origin of Coal," *Cincinnati Mirror* 4 (1835): 98–99, 104, and Benjamin Silliman, "Vegetable Origin of Coal," *Boston Cultivator* 5 (1843): 234.

22. Samuel Harries Daddow and Benjamin Bannan, *Coal, Iron, and Oil; or, the American Practical Miner* (Pottsville, Pa.: B. Bannan, 1866), pp. 63 and passim.

23. Richard C. Taylor, *Statistics of Coal: The Geographical and Geological Distribution of Mineral Combustibles and Fossil Fuel, Including also Notices and Localities of the Various Mineral Bituminous Substances Employed in Arts and Manufactures . . .* (Philadelphia: J. W. Moore, 1848). See also Walter R. Johnson, *A Report to the Navy Department of the United States, on American Coals* (Washington, D.C.: Gales and Seaton, 1844).

24. Denison Olmsted, *Observations on the Use of Anthracite Coal* (Cambridge, Mass.: Folsom, Wells, and Thurston, 1836).

25. Knapp, *Chemical Technology*, pp. 110–30; Daddow and Bannan, *Coal, Iron, and Oil*, pp. 75–81.

26. Paul H. Giddens, *Pennsylvania Petroleum 1750–1872* (Titusville: Pennsylvania Historical and Museum Commission, 1947).

27. Robert R.F. Kinghorn, *An Introduction to the Physics and Chemistry of Petroleum* (New York: John Wiley and Sons, 1983), pp. 85–92.

28. Andrew Ure, *A Dictionary of Arts, Manufactures, and Mines* (New York: Appleton, 1850), pp. 663–68.

29. F. M. Binder, *Coal Age Empire* (Harrisburg: Pennsylvania Historical and Museum Commission, 1974), pp. 28–29.

30. Ibid., p. 56. Early records are contained in: Philadelphia Gas Works, *Annual Reports of the Trustees* (Philadelphia, 1835+); Boston Gas Light Company, *Extracts from the Bye-laws of the Company, Together with Some Statements Respecting Oil Gas, Lately Received from the Liverpool Company* (Boston, 1823). See also the records of the New York Gas Light Company.

31. J. C. Booth, "Gas, Illuminating," from *The Encyclopedia of Chemistry, Practical and Theoretical: Embracing Its Applications to the Arts, Metallurgy, Mineralogy, Geology, Medicine, and Pharmacy* (Philadelphia: H. C. Baird, 1850), pp. 663–67.

32. Daddow and Bannan, *Coal, Iron, and Oil*, pp. 99–128.

33. Olmsted, *Observations on the Use of Anthracite Coal*, p. 4.

34. A good discussion of this issue appears in Lawrence A. Cremin, *American Education: The National Experience, 1783–1876* (New York: Harper and Row, 1980), pp. 1–5.

35. Benjamin Thompson, Count Rumford, to Professor Pictet, November 8, 1797, as quoted in Brown, *Benjamin Thompson*, p. 183.

36. John Lee Comstock, *Elements of Chemistry, in Which the Most Recent Discoveries in the Science Are Included and Its Doctrines Familiarly Explained* (Hartford: D. F. Robinson, 1831). The "44th edition" was issued in New York in 1842, and other printings appeared as late as 1859 (New York: Pratt, Oakley and Company). It was common practice among textbook publishers at that time to designate each new printing as an edition.

37. John Lee Comstock, *System of Natural Philosophy* (Hartford: D. F. Robinson, 1830). The "218th edition" was from 1860 (New York: Pratt, Oakley and Company), and other editions appeared as late as 1867. The National Union Catalog of pre-1956 imprints locates sixty-two different printings or editions.

38. Jane Haldimand Marcet, *Conversations on Chemistry, in Which the Elements of That Science Are Familiarly Explained and Illustrated by Experiments and Plates*, 2 vols. (London: Longman, Hurst, Rees, and Orme, 1806). The first American edition appeared the same year (2 vols; Philadelphia: J. Humphreys). This text went through at least fifteen different American editions and dozens of printings through 1857. There is some confusion as to the authorship. Marcet's name appears only in later editions, while some earlier editions credit a Mrs. (Margaret) Bryan—"Mrs. B."—as the true author. Marcet was the author of other "conversation" textbooks, including *Conversations on Natural Philosophy* (New York: Goodrich, Gilley, and Wiley, 1820).

39. Jane Haldimand Marcet, *Conversations on Chemistry* (Hartford: Oliver D. Cooke and Company, 1826), p. 39.

40. Ibid., pp. 118–19.

41. Walter R. Johnson (ed.), in Knapp, *Chemical Technology*, p. v.

42. Mrs. D. A. (Mary Johnson) Lincoln, *Boston School Kitchen Text-Book: Lessons in Cooking for the Use of Classes in Public and Industrial Schools* (Boston: Roberts Brothers, 1888), p. viii; the first edition appeared in 1883. Of special interest among the dozens of subsequent editions of this and other cookbooks is Mary Johnson Lincoln and Anna Barrows, *Home Science Cook Book* (Boston: Home Science Publishing Company, 1902 and subsequent editions).

43. Gwendolyn Wright, *Moralism and the Model Home: Domestic Architecture and Cultural Conflict in Chicago, 1873–1913* (Chicago: University of Chicago Press, 1980), pp. 153, 272–73. See also Charlotte Perkins Gilman, *The Home, Its Sphere and Influence* (Boston, 1903), and David P. Handlin, *The American Home: Architecture and Society, 1815–1915* (Boston: Little, Brown and Company, 1979), esp. pp. 401–15.

44. Rebecca Zurier, *The American Firehouse: An Architectural and Social History* (New York: Abbeville Press, 1982), pp. 192–94, 208–9.

45. Carl Bode, ed., *Midcentury America: Life in the 1850s* (Carbondale: Southern Illinois University Press, 1972), pp. 103 and passim.

46. Humphry Davy, *Elements of Chemical Philosophy* (Philadelphia: Bradford and Inskeep, 1812).

47. Samuel Griswold Goodrich, under the pseudonym Peter Parley, contributed dozens of juvenile texts on many subjects. His works relating to chemistry and fire technology include *Peter Parley's Book of Curiosities* (Boston: Lord and Holbrook, 1832 and later editions); *Cabinet of Curiosities* (Hartford, 1826); and *A Glance at the Physical Sciences* (Boston: Bradbury, Soden, and Company, 1844 and later editions).

48. Achille Cazin, *The Phenomena and Laws of Heat* (New York: Charles Scribner and Company, 1869).

49. William Thomas Brande, *Dictionary of Science, Literature, and Art* (London: Longman, Brown, Green, and Longman, 1843); see also the first American edition (New York: Harper and Brothers, 1843), plus yearly printings through 1875. Andrew Ure, *A Dictionary of Chemistry, on the Basis of Mr. Nicholson's* (Philadelphia: Robert De Silver, 1821). Ure, *Dictionary of Arts, Manufactures, and Mines* (London: Longman, Orme, Brown, Green, and Longman, 1839). The first American edition appeared in 1842 (New York: D. Appleton) and was followed by many printings and editions through the "7th edition" in 1877.

50. "To Kindle a Fire Under Water," *Scientific American* 1, no. 4 (September 18, 1845).

51. Michael Faraday, *A Course of Six Lectures on the Various Forces of Matter*, ed. William Crookes (New York: Harper and Brothers, 1868). See also a reprint: Michael Faraday, *The Chemical History of a Candle* (New York: E. P. Dutton, 1920).

52. John J. Fulton and Elizabeth H. Thomson, *Benjamin Silliman: Pathfinder in American Science* (New York: Henry Schuman, 1947), pp. 173 and passim.

53. Kirtley F. Mather and Shirley L. Mason, *A Source Book in Geology, 1400–1900* (Cambridge, Mass.: Harvard University Press, 1970), pp. 14–16.

54. William Thomson, "On the Secular Cooling of the Earth," *Transactions of the*

Royal Society of Edinburgh 23 (1862): 157–69, and "On the Age of the Sun's Heat," *Macmillan's Magazine* 5 (1862): 288–93.

55. Thomson, "Secular Cooling," p. 159.

56. Clarence King, "The Age of the Earth," *American Journal of Science* 45 (1893): 1–20.

57. Ernest Rutherford, *Radio-Activity* (Cambridge: Cambridge University Press, 1904), p. 346.

Chapter 7
Perpetuating the Flame

1. "At Home: Sweet Home," *Scientific American* 1, no. 2 (September 4, 1845).

2. Jonathan Dawson, "Coming of the Pioneer," in Marguerite Miller, ed., *Home Folks* (Marceline, Mo.: Walsworth, n.d.), 1: 8–13. Similar accounts occur in Glenda Riley, *Frontierswomen: The Iowa Experience* (Ames: Iowa State University, 1981).

3. Dawson, "Coming of the Pioneer," pp. 8–13.

4. Henry Adams, *The Education of Henry Adams: An Autobiography* (Boston and New York: Houghton Mifflin Company, 1918), pp. 10–11.

5. Samuel Goodrich, *Enterprise, Industry, and Art of Man, As Displayed in Fishing, Hunting, Commerce, Navigation, Mining, Agriculture and Manufactures* (Boston: Rand and Mann, 1849), p. 296.

6. See, for example, Ruth Schwartz Cowan, *More Work for Mother: The Ironies of Household Technology from the Open Hearth to the Microwave* (New York: Basic Books, 1983), pp. 55–56.

7. James Bryce, *The American Commonwealth*, 2 vols. (New York: Macmillan, 1912), 2: 802–3.

8. Elinore Pruitt Stewart, *Letters of a Woman Homesteader* (1914; rpt. Lincoln: University of Nebraska Press, 1961), p. 31.

9. Quoted in Elisabeth Donaghy Garrett, *At Home: The American Family 1750–1870* (New York: Harry N. Abrams, 1990), p. 101.

10. Quoted in Cowan, *More Work for Mother*, p. 56.

11. Quoted in Henry J. Kauffman, *The American Fireplace: Chimneys, Mantelpieces, Fireplaces & Accessories* (Nashville and New York: Thomas Nelson, 1972), p. 168.

12. See Putnam's treatise, *The Open Fire-Place in All Ages* (Boston, 1881).

13. "Furnishing the Nursery," *The Modern Priscilla* (May 1917): 32.

14. William Allen White, *The Autobiography of William Allen White* (New York: Macmillan, 1946), p. 34.

15. Ibid., p. 35.

16. Quoted from W. L. George, *Woman and To-Morrow* (1913), in which he also stated that "The home is the enemy of woman." Quoted in Gwendolyn Wright, *Moralism and the Model Home: Domestic Architecture and Cultural Conflict in Chicago, 1873–1913* (Chicago: University of Chicago Press, 1980), p. 157.

17. Lewis Atherton, *Main Street on the Middle Border* (Bloomington: Indiana University Press, 1984), p. 21.

18. Ann Douglas, *The Feminization of American Culture* (New York: Knopf, 1977), pp. 327–29. John Higham also discusses this shift in attitude in "The Reorientation of

American Culture in the 1890's," in *Writing American History: Essays on Modern Scholarship* (Bloomington: Indiana University Press, 1970).

19. Harold I. Sharlin, "Applications of Electricity," in Melvin Kranzberg and Carroll W. Pursell Jr., *Technology in Western Civilization: The Emergence of Modern Industrial Society, Earliest Times to 1900* (New York: Oxford University Press, 1967), 1: 563–78.

20. Luna Park Summer Concert Program, n.d. Authors' files.

21. Adams, *Education of Henry Adams*, p. 381.

22. David Swing, *The Story of the Chicago Fire* (Chicago: H. P. Gross, 1892), p. 28.

23. Firemen directed the rescue operations that followed the Planters Hotel collapse. These exploits are depicted in the Currier print.

24. For listings and discussion, see John Lowell Pratt, ed., *Currier & Ives: Chroniclers of America* (n.p.: Hammond Incorporated, 1968); Colin Simkin, ed., *Currier and Ives' America: A Panorama of the Mid-Nineteenth Century Scene* (New York: Crown Publishers, 1952); and Harry T. Peters, *Currier & Ives: Printmakers to the American People* (Garden City, N.Y.: Doubleday, Doran and Company, 1942).

25. Pollack claimed that he made more than 4,000 stereographs per day between 1872 and 1874. See William C. Darrah, *The World of Stereographs* (Gettysburg, Pa.: Darrah Publishing, 1977), p. 45.

26. See, for instance, the program dated December 23, 1913, for the Oregon Agricultural College Cadet Band. Authors' files.

27. Reprinted in Robert M. Hazen, *The Poetry of Geology* (London: George Allen and Unwin, 1982), p. 18.

28. Jesse E. Dow, "The Iron Master," *Scientific American* 1, no. 6 (October 2, 1845).

29. Harriet Beecher Stowe, *Oldtown Folks* (Cambridge, Mass.: Harvard University Press, 1966), pp. 105–8.

CHAPTER 8
EPILOGUE

1. Henry Adams, *The Education of Henry Adams: An Autobiography* (Boston and New York: Houghton Mifflin Company, 1918), pp. 466–67.

2. "A Fire Department in Action," Keystone Stereograph P152 (Meadville, Pa.: Keystone View Company, ca. 1900).

3. *People*, July 11, 1988, p. 81.

4. John W. Lyons, *Fire* (New York: Scientific American Library, 1985), p. 1.

5. Ibid., pp. 1, 4.

6. Don Oldenburg, "Playing with Fire," *Washington Post*, January 12, 1988, B5.

7. National Fire Prevention and Control Administration, *Arson: America's Malignant Crime* (Washington, D.C.: Government Printing Office, 1976), p. 5; *Sourcebook of Criminal Justice Statistics* (Washington, D.C.: Government Printing Office, 1987), p. 291.

8. Russell Lynes, *The Domesticated Americans* (New York: Harper and Row, 1957), p. 263. For an excellent study of the electrification of America, see David E. Nye, *Electrifying America: Social Meanings of a New Technology, 1880–1940* (Cambridge, Mass.: MIT Press, 1991).

9. Oldenburg, "Playing with Fire."

10. In the early days, televisions were even built into bricked-up fireplaces. For more on this, see Bob Shanks, *The Cool Fire: How to Make It in Television* (New York: Norton, 1976).

11. Quoted in Ruth Schwartz Cowan, *More Work for Mother: The Ironies of Household Technology from the Open Hearth to the Microwave* (New York: Basic Books, 1983), p. 56.

Index

absolute zero, 188
acid rain, 240–241
Adams, Andy, storyteller, 63
Adams, Henry: on electricity, 224; on fire
technology, 16; on home fires, 217; on
progress, 236; at St. Louis Exposition,
15
advertising: boilers, 27; fireplaces, 20; lighting devices, 22, 87; stoves, 7, 16, 162,
166–167, 169, 173; use of fire in, 5, 21,
54–57
Aetna Insurance Company, 133
Alabama: Mobile fire laws, 113; Mobile
fire of 1827, 119; Shiloh Baptist
Church, 65
alarms: telegraphic, 124–125; wooden rat-
tles, 125
alchemy, 183
Alcott, Louisa May, Little Women, 107,
154, 251
Allen, Lewis F., on fireplace design, 219
Allison, Emma, 154
Allston, Washington, 62
American Fire King. See Houghton, W. C.
American Frugal Housewife, 20, 109
American Museum: destroyed by fire, 80,
151; fire of 1868, 37–38, 152
anthracite coal: acceptance in United
States, 198; definition, 196; proper use,
115, 197–198; stoves, 164
Anvil Chorus, use of firemen in, 143–144,
260
Argand, Ami, inventor of lamp chimneys,
174
Aristotle, 12
Arizona: firefighting, 129; need for fire,
106; Tombstone fire, 75
arson, 99–106, 240; attitudes toward, 256;
investigation, 103–104; motivations,

105–106; techniques, 103; use against
blacks, 101, 102, 104, 127
art: depictions of the fireside, 230; fire imagery, 224–230; panoramas, 225
asbestos: definition, 258; in fireproofing,
117
ashes: disposal, 161; uses, 19, 21–22, 23, 25
assaying, 29
atoms, vibrations of as heat, 188–189
automobile, 236

Baker, Benjamin A., A Glance at New
York, 147–148
Ball, Nehemiah, popular lectures, 210
balloons, hot air, 26, 247
Baltimore: firefighting, 127–128; fire of
1904, 74, 254; gas lighting, 30, 201
bands: in firemen's parades and gatherings,
144–145; programmatic fire music, 230–
231; Rhode Island, 47; serenading by,
44
Banvard, Henry, panoramic painting, 225
Barnard, Hannah, burn treatments, 94
Barnum, Phineas T.: fire losses, 80, 151,
152; museum fire, 37; and Phillips Anni-
hilator, 55, 110–111
Barnwell, Edward, and fire magic, 41
Beard, Daniel: American Boys Handy Book,
49–52; fire accident, 52
Beecher, Catharine: American Woman's
Home, 92; on fire prevention, 115; on
importance of the hearth, 243
Bennett, Mary, on starting a fire, 154
Berger, Lydia, arsonist, 105
Biddle, Ellen McGowen, fire experiences,
106
Bierstadt, Albert, and fire imagery, 224
bituminous coal: definition, 196; smokiness
of, 82–83